The

NORTH
CASCADES
HIGHWAY

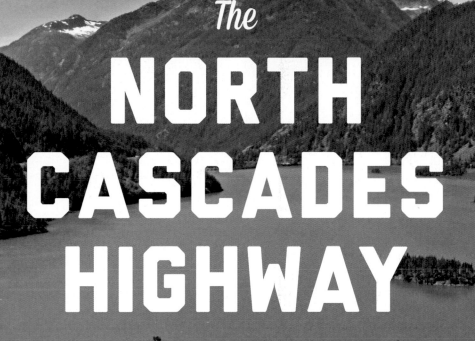

The
NORTH
CASCADES
HIGHWAY

A Roadside Guide to America's Alps

JACK McLEOD

UNIVERSITY OF WASHINGTON PRESS

Seattle and London

August wildflower (sketch by Kira Violette)

© 2013 by the University of Washington Press
Printed and bound in the United States of America
Design by Thomas Eykemans and Molly McLeod
Composed in Chaparral, typeface designed by Carol Twombly
Display type set in Liberator, designed by Ryan Clark
16 15 14 13   5 4 3 2 1

UNIVERSITY OF WASHINGTON PRESS
PO Box 50096, Seattle, WA 98145, USA
www.washington.edu/uwpress

Library of Congress Cataloging-in-Publication Data is on file
ISBN 978-0-295-99316-4

The paper used in this publication is acid-free and meets the mini-
mum requirements of American National Standard for Information
Sciences—Permanence of Paper for Printed Library Materials, ANSI
Z39.48–1984.∞

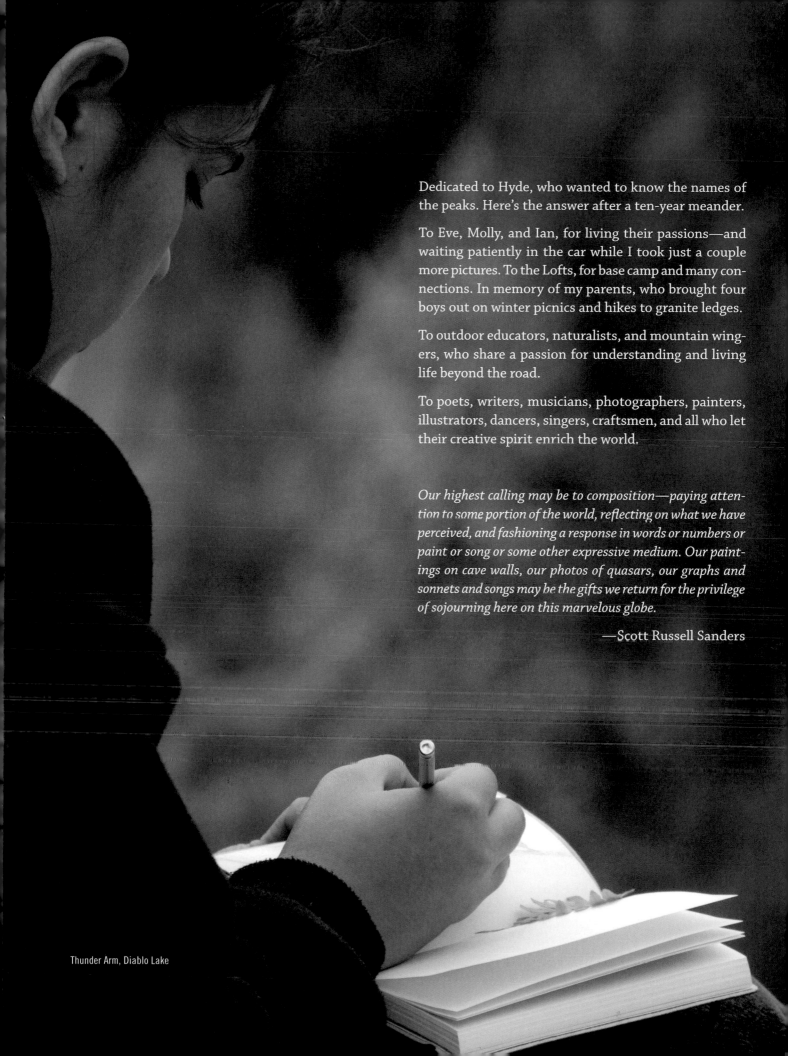

Dedicated to Hyde, who wanted to know the names of the peaks. Here's the answer after a ten-year meander.

To Eve, Molly, and Ian, for living their passions—and waiting patiently in the car while I took just a couple more pictures. To the Lofts, for base camp and many connections. In memory of my parents, who brought four boys out on winter picnics and hikes to granite ledges.

To outdoor educators, naturalists, and mountain wingers, who share a passion for understanding and living life beyond the road.

To poets, writers, musicians, photographers, painters, illustrators, dancers, singers, craftsmen, and all who let their creative spirit enrich the world.

Our highest calling may be to composition—paying attention to some portion of the world, reflecting on what we have perceived, and fashioning a response in words or numbers or paint or song or some other expressive medium. Our paintings on cave walls, our photos of quasars, our graphs and sonnets and songs may be the gifts we return for the privilege of sojourning here on this marvelous globe.

—Scott Russell Sanders

Thunder Arm, Diablo Lake

Wildflowers, Cutthroat Basin. Liberty Bell group, upper left. Blue Lake Peak, upper right.

Contents

Davis
Peak

Mount Ross

Sourdoug
Mountai

GORGE LAKE

Mount Watson

*DIABLO
LAKE*

DIABLO

NEWHALEM

Colonial Peak

THUNDER CREEK

Sauk Mountain

MARBLEMOUNT

ROCKPORT
MILE 97.6

SKAGIT RIVER

Eldorado Peak

Cascade
Pass

THE NORTH CASCADES HIGHWAY

WASHINGTON
STATE

N

0   5 MI

0   10 KM

ROSS LAKE

Jack Mountain

Crater Mountain

Ruby
ountain

Granite Creek Canyon

METHOW RIVER

McLeod
Mountain

Graybeard
Peak

The Needles

Porcupine
Peak

MAZAMA

MILE 180

Cutthroat Peak

Silver Star Mountain

Rainy Pass

Washington Pass

Early
Winters
Spires

Stiletto Peak

McGregor
Mountain

August evening on Sourdough Mountain

Invitation

How can I give directions to a time, to a place where magic happens?
—Saul Weisberg, executive director, North Cascades Institute

HIGH ON THE RIM of Sourdough Mountain, evening light replenishes my soul. This is why we visit the North Cascades. To slow down, to decompress, to revive. I watch summer's glow illuminate sepia cliffs and a kaleidoscope of blossoms while across the valley, steep, snow-covered pinnacles soften in warm pastels. The tiny road I left five thousand feet below twists around turquoise Diablo Lake. Alpine fragrances hang in the air—sun-baked fir mingles with ephemeral aromas of wildflowers and the earthen smell of sixty-nine-million-year-old Skagit Gneiss trail dust. I've arrived. I've walked into a miraculous convergence of August sun, a thousand blooms, and a North Cascades ridge looking over the world.

I first crossed the North Cascades Highway years before with my wife, an extra treat at the end of our honeymoon. Curious about what was beyond this spectacular road, we took a two-mile hike to Lake Ann and were quickly immersed in a world with woodlands and alpine wildflowers. Beyond the forest was a cliff-ringed jewel of a lake and views to savor for weeks and years to come—unexpected treasure for stopping our car and taking a simple walk in the woods.

Many visits later I've found treasure awaits everywhere just off the asphalt. Moss-encased rocks lie silently in fern-filled forests. Amphitheaters of peaks and spires abruptly rise above the road, the earth's exoskeleton exposed in geologic splendor. Colonies of carnivorous plants await their prey next to still ponds. Quiet streams gurgle under ancient cedars. Waterfalls roar through granite cracks, delivering ice melt from unseen glaciers above. Ravens play in wind currents over mountain-rimmed lakes. And everywhere, the intoxicating mountain air is a prescription for lower blood pressure.

Thousands of drivers cross the North Cascades Highway admiring the views but not knowing the place. Alas, travelers with their eyes on the end of the journey don't miss what they don't know they're missing. This book was conceived when one person asked one simple question. She wanted to know the names of the peaks. Names bring familiarity. Familiarity brings curiosity to know more. And soon we appreciate the deeper beauty and our connections to a world older and more intricate than we had imagined.

The North Cascades Highway leads through a vertical world extraordinary in beauty and access. It took months for early explorers to cross the North Cascades, and now it can be done between breakfast and lunch. But seeing these sights by car is a recent privilege. In

Twilight, Rainy Lake

the 1890s, when Thomas Edison was designing light-bulbs, Washington State established a road commission to search for a route across these rugged mountains. Those early commissioners could not have dreamed that men would walk on the moon before their road was built. Survey parties explored all possible passes, but steep canyons, deep snows, and economics blocked their efforts. Repeated attempts and increased pressure for cross-state commerce finally resulted in the 1972 opening of the North Cascades Highway eleven years after humans traveled to space.

This book is based on a highway, but it is about what's on the sides. The familiar is the key to unraveling the unfamiliar. On this, one of America's most scenic highways, the beautiful and spectacular are hard to miss. Through photography and short essays, I hope the richness of these landscapes becomes more visible.

The realm beyond the highway is full of hidden delights—botanical, zoological, geological. This book will help you unearth the tales of once-molten rock and remnants of ocean floor now towering above the road. Take this book into the North Cascades and take your time. A stroll off the road is transportation to a tranquil world of ancient sounds and smells—a time machine—to a world that begins where the pavement stops. Become familiar with the place. Look more closely and appreciate the beauty at a pace not measured in miles per hour. Pause, admire the views, and create your own stories.

The book's organizing principle is road mileage, from west to east, from mile 97.6 (near Rockport) to mile 180 (Mazama). The geologic stories may seem without a continuous story line, because unlike the Grand Canyon and other parts of the country, North Cascades rocks don't transition from oldest to youngest. Jigsaw puzzle or quilt analogies come up frequently. There is narrative, though. There are ancient processes here, and time beyond comprehension. Each mileage stop tells one piece of the North Cascades story.

Pyramid Peak, sunrise, March

Acknowledgments

I'm grateful to the following people for helping make this book happen.

Molly McLeod, who does graphic design like most of us breathe.

Writers who inspired this high school literary magazine reject to try again decades later: John McPhee, William Dietrich, Timothy Egan.

Geologists who clarify the complicated North Cascades story: Ralph Haugerud, Roland Tabor, Jon Reidel, Scott Babcock, David Tucker, Al Friedman, Brad Smith.

Bob Mierendorf and Jesse Kennedy, for bringing the past to light.

Char Seawell and Marie Doyle, for reading and encouraging.

Ana Maria Spagna for a key tip.

Lofty, for beta-testing on the road.

Saul Weisberg, who wanted people to know the place and built a legacy at the North Cascades Institute. And all the staff and grad students who put heart, soul, and mind into educating and nourishing the gifts of others. You do inspire close relationships with the natural world.

Bonita Hurd, who is a master of detail and copy editor extraordinaire.

I would like to thank teachers who modeled an eager love for discovery, setting me on a path of lifelong learning. Mr. Bonner (fourth grade), who provided a feast of predicting and problem-solving. Mrs. Dazé and the basement team in Miller School (sixth grade), who constructed a world of integrated, project-based learning. Mr. Yaggi (eighth grade), who captained a science extravaganza. Dr. Cappel (ninth grade), who taught fourteen-year-olds to meticulously observe and persevere. Dr. Romey and Dr. Elberty, SLU, who frequently removed boundaries to learning.

Large-leaved lupine (*Lupinus polyphyllus*)

Many others who pushed, led, and let us trek into the unknown to see what we would find. And all those who still do. Teachers change the world. (Full disclosure: I joined their ranks. This is for my colleagues around the world.)

And Eve, who still likes to pick up pebbles.

The
NORTH
CASCADES
HIGHWAY

Sourdough Mountain lookout

From Miners to Poets

The spectacular scenery and raw wildness are only the outer wrapping that packages a story of bygone people and ancient landscapes.—Bob Mierendorf, National Park Service archeologist

HUMAN HISTORY in the North Cascades is rooted in geology. Ancient travel routes followed valleys or ridges formed and carved by tectonic and glacial processes. Prehistoric explorers searched for animal and plant resources but geologic treasures were also a lure. As early as eight thousand years ago, chert—quartz formed from ancient microscopic organisms that fractures like obsidian and so can be flaked into sharp-edged pieces—was mined for making scrapers, knives, and other tools. Hozomeen Mountain, towering above some of these ancient quarries, gets its name from a Native American term for "sharp like a knife." Historic tools made from North Cascades chert have been found in various parts of Washington, evidence of trade between early cultures. The first Cascades miners were not 1800s gold-seekers but the native Skagit people.

Eight millennia later, gold drew prospectors up the Skagit River to Ruby, Thunder, and Slate Creeks. The 1880s brought thousands of miners and hundreds of claims, but since only a few of them made money, most left almost as soon as they came. Some stayed, building cabins and living off the land, trying to find their fortune in the hard rock of the North Cascades. Few physical artifacts remain, but hints of their stories linger in the landscape's names: Jack, Ruby, and Sourdough Mountains; Hidden Hand Pass, Roland Point; May Creek.

Sporadic explorers came to hunt, trap, log, homestead, or find routes across the mountains. Scottish fur trapper Alexander Ross crossed in 1814 with three native guides (Ross Dam and Mountain were named after J. D. Ross, leader of the dam projects). Some adventurers sought the high peaks and granite cliffs just for the challenge and thrill of getting to the top. Climbers pioneered routes in the deepest parts of the

mountains forty years before the highway existed, and peak names testify to their challenges: Terror, Forbidden, Torment, Triumph. If you look carefully, you may see modern-day climbers ascending thousand-foot cliffs at Washington Pass.

During the 1950s, notable poets and writers also made it to North Cascades mountaintops. Jack Kerouac and other influential members of the Beat Generation spent solitary summers writing and reflecting while manning the fire towers on Desolation Peak and Sourdough and Crater Mountains.

Today, people are drawn to the soul-inspiring North Cascades to walk, climb, write, sketch, camp, canoe, bike, ski, take photographs, breathe the mountain air, or simply be still and hear the landscape speak. We all come for the unspoiled beauty, which still exists because dedicated people thought it was important enough to protect, geology and all.

Chert scraping tool

Remains of a ranger station

Finding peace in the alpine

Gneiss

Subterranean forces are evident in the rock all along the highway.

A Shuffled Deck of Rocks

Terrane: A rock formation or assemblage of rock formations that share a common geologic history. An exotic terrane is one that has been transported into its present setting from some distance.
—United States Geological Survey

WASHINGTON is a collection of geologic immigrants. Lithologic drifters. A crazy quilt of igneous, sedimentary, and metamorphic rocks from afar. Minicontinent-size land masses and oceanic terranes have been moving in, taking up residence and permanently adding real estate on the western margin of North America for hundreds of millions of years.

There didn't used to *be* a Washington. The continental coastline was near the border of Idaho a hundred million years ago but seafloor sediments and volcanic islands kept colliding with the west coast, driven by the slow but colossal force of tectonic plates. Land was added. Mountain ranges from Northern California to Alaska rose up, eroded to high plains, and rose again.

Like people, immigrations came from different places and times. Some residents kept their native looks—Hozomeen pillow basalts still look like underwater lava flows (except they're atop a mountain). The origins of other newcomers are not so clear. Shuksan Greenschist and rocks in the Skagit Gneiss Complex don't look like their ocean-bottom origins because they've been recrystallized by heat and pressure, deformed, and reconstructed into new forms and compilations.

Some geologic settlers have resided here for so long they've had progeny, eroding and depositing elsewhere to form new rocks such as Winthrop Sandstone. More recently, volcanoes (Mount Baker and Glacier Peak) have

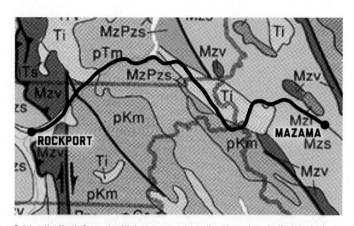

Driving the North Cascades Highway means tunneling through a shuffled deck of rocks and timescales. Each map color and label represents different groups and ages of rocks.

risen from the depths and taken over their own pieces of real estate. All these geologically diverse neighbors create geologically diverse neighborhoods rich in history and stories.

GEOLOGIC JIGSAW

Geology is about uncovering the stories, which in Washington means piecing together the parts of the puzzle. It begins with careful observation. Notice colors and textures in a roadside cliff. These are clues. You may be looking at segments of a city-sized mass of magma that wedged, baked, and buckled surrounding rock as it rose through the crust ninety million years ago. It could be millions of years older than its geologic neighbor. At fifty miles per hour, pass fifty million years in a blink.

The North Cascades are a mixed collection of rocks. Underlying the beauty, underlying the trees and peaks and lakes, underlying our first gasps of wonder at this magnificent landscape are geologic stories that expand our concepts of time and our place on earth.

Pond lilies (*Nuphar polysepalum*), Thunder Lake

Spiny wood fern
(*Dryopteris expansa*)

False hellebore
(*Veratrum viride*)

White-tailed ptarmigan
(*Lagopus leucura*)

Pinesap
(*Hypopitys monotropa*)

Western trillium
(*Trillium ovatum*)

A Hidden Realm

You need to get on your feet to really experience this place.
—Jon Riedel, National Park Service geologist

IN CHILDREN'S FICTION, characters walk through wardrobes and walls into parallel worlds with exotic creatures, giant trees, and carnivorous plants. Unforeseen adventures ensue, often humbling and enlightening the visiting human. Such a parallel world exists. Here.

Views through the window are impressive, but our moving car separates us from the world outside. Removing the veils of glass and speed, we gain entry to a lavish realm. Open the car door. Walk a bit. Pause. Our pace slows, but our synapses supercharge as diverse parts of our brain awake. Sensory portals open. Mountain scents appear in cornucopian abundance; woodland textures intrigue, soothe, or harshly provoke us (don't touch the devil's club!); sounds reveal unseen activity: a clattering rock, a trickle of water, the calls of the varied thrush (which isn't varied) and Swainson's thrush (which is).

I sit. Time slows. My senses waken. I'm glad for this dry patch under the giant red cedar. I finally notice vertical threads of electric-green moss. After-rain drips whisper somewhere in the mist. My hand rests on a thick blanket of stair-step moss that cushions a mound of granite boulders. A subtly fragrant cocktail of damp decay, cedar bark, and crisp air lingers in the fog.

But this realm is secret. And unknown and unknowable if we don't take time to pause.

Even with patient observation, much remains hidden in the North Cascades. More than four hundred species of mammals, reptiles, amphibians, birds, and fish. The secretive wolverine and lynx. And *U. arc-*

tos horribilis has returned, the grizzly's iconic hump silhouetted on a ridgetop. Thousands of plant species here inhabit eight distinct life zones with different combinations of elevation, climate, slope, exposure, and geology. Even with abundance, it takes patience, awareness, and some luck to observe this diversity of life. One June day, a group of naturalists heard or saw more than sixty species of birds between the Cascades crest and the Methow Valley. What do we see at fifty miles per hour?

HIGHWAY PARADOX

The North Cascades Highway was designed to unobtrusively thread through this thousands-of-years-old wilderness. Drivers benefit by seeing the land unmarred. Narrow as it is, though, this portal to the wilderness is also the single biggest scar and intrusion, bisecting wildlands, often traveled by more than a thousand vehicles a day.

The highway affords access to what it defaces, but the DNA of this landscape remains. From woodland orchids to alpine glaciers to the edge of a pond where whole worlds of miniature plants and animals colonize a single floating log. From the tiny red lipstick cladonia lichen to groves of ancient cedars. The hidden realm is abundant.

The road is a transparent tunnel guiding travelers efficiently from end to end. The highway weaves through majestic beauty but is separate from it. Beyond the glass is an extraordinary realm, understood best on foot.

Mile 134. Colonial and Pyramid Peaks

Mile 154. Black, Repulse, Fisher Peaks

Mile 162. Washington Pass

Mile 165. Early Winters Spires and Liberty Bell Mountain

Driving and Viewing Tips

Traveling "the most scenic mountain drive in Washington," it is easy to get distracted. Watch for bicyclists and pedestrians, and use roadside turnouts to stop and safely admire the views.

THE NAVIGATION SYMBOLS at the page tops will let you know in advance if you'll be able to park and from which direction(s) the views can be seen. The yellow star on each page indicates the described area's location along the North Cascades Highway. A circled P means there is a safe place to pull off and park. A car facing east or west indicates the driving direction from which the photo view can be seen.

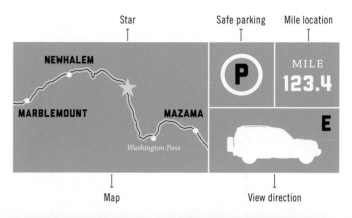

Star · Safe parking · Mile location

NEWHALEM

MARBLEMOUNT · MAZAMA

Washington Pass

P · MILE **123.4** · **E**

Map · View direction

Safely admiring the views in a parking pullout. Mile 125, Gorge Lake.

Dorado Needle
(8,440 feet)

Eldorado Peak
(8,886 feet)

The "knife edge" on top of Eldorado Peak in the Chelan Mountains Terrane of the Metamorphic Core Domain

Ninety-million-year-old Eldorado Orthogneiss

Two-hundred-million-year-old Sauk Mountain submarine volcanic breccia

NEWHALEM
MARBLEMOUNT
MAZAMA
Washington Pass
P
MILE
100
E

Eldorado Peak 8,868 FEET

A majestic view into the high peaks waits at milepost 100. Rising almost nine thousand feet, Eldorado Peak is the apex of three ridges that include about twenty summits over seven thousand feet. First summited in 1933, Eldorado Peak is a popular destination for experienced climbers who hike more than fourteen thousand vertical feet from car to knife-edge peak and back. One of the glaciers they cross is appropriately named Inspiration. But the mountain, born of searing magma, will not last; ice brings decay.

ELDORADO IS COMING DOWN. Ice and water are conspiring with gravity to wear Eldorado and surrounding peaks to a more pedestrian-friendly elevation. The Inspiration, Klawatti, and Eldorado Glaciers transfer entombed rocks to precipitously steep creeks, where stones crash and smash into ever-changing collections of fragments. Far, far downstream from Eldorado's alpine heights, the Skagit River deposits specks of Eldorado Peak at your feet. Pick up a fistful of sand and hold the North Cascades in your hand.

ALMOST A HUNDRED MILLION YEARS IN THE MAKING
Eldorado Peak's story stretches back ninety two million years. Long before the Cascades were here, a city-size body of molten rock slowly cooled deep underground, the roots of an earlier mountain range. In time, land masses and seafloor sluggishly bulldozed their way into the continent, subducting, squeezing, fracturing. In some places, rock melted. In others, pressure and heat separated light and dark minerals to form a color-banded metamorphic rock called orthogneiss. Almost fifty million years later, forces renewed, uplifting the landscape and propelling Eldorado Peak skyward, higher even than what you see today. Now, glaciers and gravity sharpen the peaks' outlines while ultimately reducing them to insignificance.

DOMAINS AND TERRANES
Lean on the park fence at mile 100. It's peaceful. The Skagit endlessly flows. Eagles and Eldorado Peak rise

Volcanic remains from the ocean deep, Sauk Mountain is part of the Chilliwack River Terrane. In the background, Mount Baker is part of the Nooksack Terrane. Both are in the Western Domain.

in the distance. Ah, life in the Chilliwack River Terrane, Western Domain. Gazing at Eldorado is the equivalent of looking out your window to a different part of history and geography. Eldorado is in the Metamorphic Core Domain's Chelan Mountains Terrane. North Cascades mountains are so different in age, origin, and rock type that geologists have divided the region into three major, interweaving geologic domains, each further divided into two to four terranes (see map, page 75).

You, and Sauk Mountain behind you, are in the Chilliwack River Terrane, composed of altered seafloor sediments and lava from an ancient volcanic island chain. Eldorado began as a mass of magma in a different time and place. Farther east is the Methow Domain with its own history. A crazy quilt is orderly by comparison.

The Skagit
Lifeblood of Salmon and Eagle

Where better than such a place to recognize that the essence of nature is flow—
of lava, electrons, water, wind, breath.—Scott Russell Sanders

WHEN THE NORTH CASCADES HIGHWAY was nearing completion in 1972, bald eagles were nearing extinction. Now, because of habitat protection and bans on the pesticide DDT, you can once again see eagles on gravel bars, in the trees, and in the air. This stretch of river is part of the 2,450-acre Skagit River Bald Eagle Natural Area, a piece of the larger Skagit Wild and Scenic River System, and is a popular winter location to see eagles congregating and feeding on spawning salmon. My favorite viewing spot—right next to the road—looks down on a river bend where clusters of young and mature eagles jockey for position at the salmon sushi bar. A record 580 eagles were counted one January day along an 11.5-mile stretch of this river. In late winter, the majestic birds leave with their lifelong mates for nests in Alaska and British Columbia. A few remain in the area, sometimes visible along the river or gliding in the sky above.

The Skagit has always been this region's artery. Ocean-going salmon swim back upriver to spawn, delivering life back to stream and forest. Early hunters, traders, trappers, hopeful miners, and even tourists paddled by dugout canoe or paddlewheel steamer as far as the river could be navigated (present-day Newhalem). Even mountains travel the river as eroded fragments, endless freight on a one-way trip to the ocean, pausing only to become nutrient-rich flood deposits in Skagit Flats farmland.

With headwaters in Canada's Manning Provincial Park, and with tendrils reaching to Snohomish County's 10,541-foot Glacier Peak and Whatcom County's 10,778-foot Mount Baker, the Skagit system drains more than twenty-nine hundred streams that feed ten billion gallons of freshwater a day into the salty Salish Sea.

The Skagit River supports healthy populations of all five species of native salmon—chinook, chum, coho, pink, sockeye—and cutthroat and steelhead trout. This king (chinook) is spawning in a tributary stream.

Even as they die, salmon give life back to the forest, providing nitrogen and other nutrients.

A bald eagle (*Haliaeetus leucocephalus*) with an impressive wingspan glides above the Skagit River.

Carefully balanced Skagit Gneiss will remain until the river rises.

A fly fisherman casts in the peaceful dusk.

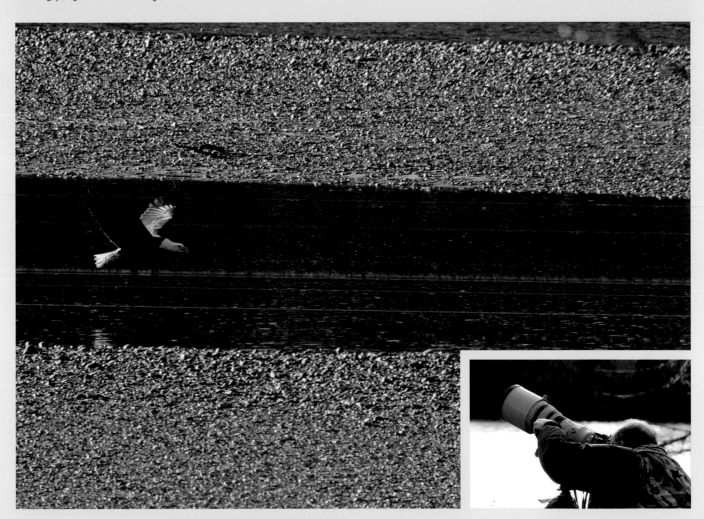

Photographers throng the Rockport area during January's Skagit River Salmon Festival.

In the distance, Mount Ross (formerly Mount Goodell) rises more than a mile above the sediment-filled valley. October snow blankets higher elevations in white.

Post-ice-age sand and gravel deposits (Quaternary alluvium) fill the valley.

Diobsud Creek crosses the Straight Creek Fault at mile 108.6. Diobsud Butte (5,858 feet) rises in the background.

Skagit Valley: Tales Below

The first rampart of high roadside peaks, 6,052-foot Mount Ross rises more than a mile above the hamlet of Newhalem, seventeen miles up the road. The Skagit River has meandered through and widened this valley for tens of thousands of years, but the story here is underneath—the greatest events are sometimes hidden from view. Two geologic diaries of ancient encounters loiter here: a tectonic fault and millions of cubic feet of glacier-delivered rock. They are memoirs stored in a geologic basement—old, concealed archives of what came before.

CONCEALED DIARIES OF ICE . . .

ICY INVADERS once inched from their hideouts in the hills, bringing riches down to the valley. You're driving across their remains. Glaciers are geologic escalators with one direction: down. Their passengers? Rocks and sediment. Creating a reversal in topography, these fragment-filled tributaries of ice scrape at alpine heights then transport their captive stones to valleys far below. About fifteen thousand years ago they spilled out like fifteen million dump trucks leaving their loads. As the last ice age ended, the glaciers melted away, leaving behind deep deposits of sand and gravel—former mountaintops relocated to future floodplains. Excavated all across the northern United States, these glacial deposits are now used as a resource to make roads, like the one transporting you back up the glacier's path.

. . . AND COVERT CONTINENTS

If glaciers are escalators, earth's crustal plates are planetary barges transporting geologic cargo thousands of miles. A continental tryst occurred here: crustal ships passed in an eons-long night, leaving a new landscape and buried faults.

Two tectonic plates slowly slid past each other like today's San Andreas fault system in California, where the Pacific Plate slides northward relative to the North American Plate. Hidden under this Cascades valley is the Straight Creek Fault, an ancient analog to the San Andreas. Until about forty-five million years ago, land slowly slid north for tens of millions of years. This migrant landscape traveled from somewhere south, delivering the ocean sediments and volcanic rocks that became the hills west of this valley. Eastward are the high peaks of the North Cascades—a geologic zone torn and twisted by the pressure of earlier continental collisions.

MEGATHRUST EARTHQUAKES

Earthquake danger from the Straight Creek Fault is long past, but earthquakes still occur in western Washington owing to another geologic escalator. A once-mighty slab of ocean crust, the Juan de Fuca Plate is slowly descending and being consumed under the North American Plate. The stress caused by this collision creates earthquakes, like the catastrophic magnitude-nine earthquake that struck western Washington on January 26, 1700. (The precise date is known from detailed records of a resulting tsunami that struck Japan.) Seismologists have determined that similar "megathrust" earthquakes occur in the Pacific Northwest every three hundred to five hundred years.

The story is not over, nor will it ever be. The Juan de Fuca Plate will be overtaken. The landscape reshapes as climate sculpts the surface and tectonic forces work below. Both are transforming today's temporary scene into an unrecognizable future.

Cascade River Road and Cascade Pass

At mile 106, across the bridge, then drive twenty-three miles on a narrow road, sometimes paved, sometimes dirt.

JAW-DROPPING VIEWS of alpine glaciers and their carved-up domain greet you at the end of the road. Cascade Pass was an ancient trade route used by indigenous travelers to cross between the lush, green forests of the Skagit River valley and the dry plains of the Columbia Plateau. A nearby archaeological site with stone tools was dated to ninety-six hundred years ago. The highway was originally planned to cross here, but the pass was too steep and avalanche prone.

Easy access and spectacular panoramas make this the most popular day hike in North Cascades National Park. The Marblemount ranger station provides visitors with timely information about trail conditions and bear sightings.

Above Cascade Pass, spectacular views look east to Stehekin Valley and future adventures.

A hoary marmot (*Marmote caligata*) strides past an alpine camp.

Hydrothermal intrusions left mineral deposits at Cascade Pass that attracted gold and silver prospectors eighteen million years later.

The glaciers of 8,200-foot Johannesburg Mountain, visible from the parking lot, are famous for prompting visitors to stop and look as chunks of ice calve off and rumble down the mountain.

Cascade Pass is often cloudy on the west side but clear on the east. Marine air moves in as a silent sea of clouds flooding from the west and filling the valleys, leaving archipelagos of peaks, as in this sunrise view of the 7,240-foot Triplets (seen from Sahale Glacier above Cascade Pass). Cascade Pass is a juncture of tectonically induced faults and molten intrusions: ocean sediments and volcanic rocks baked and compressed into metaconglomerate, schist breccia, orthogneiss, and other metamorphic rocks. The Triplets are composed of shattered schist cemented with quartz.

October color blankets marmot territory.

Cascade Pass hikers, with Eldorado Peak in the background

Leave the flatlands behind. Skagit Gorge is the beginning of a sixty-mile mountain journey.

Fifty-ton landslide boulder

Aerial photo of 2003 landslide that dropped a hundred tons of rock on the highway

Skagit Gorge

One turn is all it takes. Leave behind the flatlands and civilization for a world of steep, gravity-sculpted walls, curves, and wilderness. The North Cascades couldn't have an entrance more abrupt than the Skagit Gorge, a steep chasm cleaving a vertical mile below Mount Ross. Until the road opened in 1968, this was the major barrier to entering the mountains beyond. The road is now navigable by car, but there are still no restaurants, shops, or traffic lights for the next sixty miles.

TURN THE CORNER and a simple sign warns ROCKS. Geologic processes are still at work eroding the canyon walls, dropping a random rock here or occasional landslide there. The white streak in the large photo is the scar from a major landslide one November that released more than a hundred tons of rock and trapped residents up the road in the hamlet of Diablo. Record amounts of rainfall had occurred the previous month, saturating and adding weight to the soil. Nightly freeze-thaw cycles had weakened the unstable rock, causing the early-morning slide. The hillsides are now regularly monitored, including by electronic geophones installed by the U.S. Geological Survey.

THE GRAND OBSTACLE

The scale of this canyon—one of the deepest in North America—requires an adjustment in perspective. Unseen beyond the canyon heights, Mount Ross rises almost six thousand feet above the town of Newhalem (elevation 515 feet). In the 1880s, miners built dangerous elevated trails along the cliffs or journeyed all the way around through Canada to avoid this route. The infamous Goat Trail included sections through the gorge called Devil's Elbow (below the tunnel at mile 122.5), Wilson's Creep Hole, and Frightful Chasm. Prospectors had to navigate a series of ladders up and over the rock rib of Devil's Elbow. The 1896 road commission surveyors declared the road should go elsewhere because this route was not practical.

WATERFALLS, AN ICE DAM, AND A BREACHED DIVIDE

It's all skinny and tall here. Steep cliffs and narrow clefts with water rushing through every crevice. When the winter snowpack melts, torrents flow down vertical gulleys producing a gallery of waterfalls along the road.

One of the most dramatic is Gorge Creek Falls, at mile 123.4. Park, walk out on the bridge, but don't look down through the grating if you're afraid of heights. The waterfall drainage splits Mount Ross on your left (west) and Davis Peak on your right. Both mountains are built of twisted Mesozoic Skagit Gneiss. Not visible from the road, the precipitous gorge continues above the falls for thousands of feet, slicing the mountains in two.

The Upper Skagit River didn't used to drain this way. A divide existed here that forced river water from what is now Diablo and Ross Lake basins to drain north into the Fraser River valley. The Skagit gorge likely formed after ice sheets dammed the Skagit River thousands of years ago and then flooded, breaching this divide, eroding this canyon, and altering the course of the river from north to west. This same process released the great Lake Missoula, scouring out the scablands of eastern Washington and the bowl for Grand Coulee Dam.

Gorge Creek Falls

The powerhouse near mile 121 has four generators producing 159,000 kilowatts of electric power. Behind this building are trails, one leading to steep Ladder Creek Falls. In the early 1930s, J. D. Ross transformed the trails with exotic flowers, goldfish ponds, and light shows attracting nearly two thousand people a week to tour the area and see the dams. Near Diablo, a small zoo included peacocks, African lovebirds, albino deer, and Mexican black squirrels.

Ross Dam (540 feet): the waffle design was meant to eventually help build "High Ross" dam, which never happened.

Diablo Dam (389 feet): completed in 1930, it was the highest dam in the world at that time; it became fully operational in 1937.

Gorge Dam (300 feet) diverts water through a two-mile tunnel to the power-house in Newhalem. It is viewed here from mile 123.

Three Dams: Gorge, Diablo, Ross

AMERICA WAS INDUSTRIALIZING. The first moving assembly lines churned out mass-produced Model T's for consumers and trucks, tanks, and weapons for World War I. It was the turbulent 1910s. The United States became the most productive economy in the world, and demand for electricity was critical. J. D. Ross, superintendent of Seattle City Light (a public utility), wanted to produce electricity by building dams on the Skagit River, but a Boston company held most of the permits. When the permits expired, Ross lobbied federal officials, declaring that public rather than private operation of hydroelectric dams would best serve the public interest. In December 1917, Seattle City Light received a permit to build on Ruby Creek.

The Ruby Creek site became too expensive, but in 1919, construction began on Gorge Creek on a twenty-five-foot wooden crib dam. The dam and tunnel project faced delays because of hard granite, snow, mudslides, floods, a forest fire, the discovery of gold nearby, and a fifteen-day labor strike. The first generator was finally installed in 1924. The current three-hundred-foot Gorge Dam was finished in 1961.

Diablo Dam construction began in 1927, but political, economic, and technical issues concerning the generators and powerhouse delayed full operation until 1937. Once the dam was online, Seattle electric rates plunged to less than a penny per kilowatt-hour. Diablo's 1930s art-deco lampposts still line the narrow, curvy road crossing the dam.

Last completed was Ruby Dam. Renamed after J. D. Ross, the dam was finished in 1956, sixteen years before the opening of the North Cascades Highway. Its four generators produce more than half the 695,000 kilowatt output of the three dams, which together supply more than 90 percent of Seattle's electricity.

Gorge crew, 1935

Ross Dam flooded twenty-four miles of valley and covered more than five thousand acres of Canadian land, prompting lawsuits and mitigation. Plans to raise the dam another 125 feet met such resistance that Skagit dam building came to an end.

Northwest dams are known for destroying salmon runs. Salmon never ran the upper Skagit, however, because the steep Skagit Gorge was already a natural barrier. But changes in downstream water flow required Seattle City Light to fund creation of the Marblemount fish hatchery. More recent water-quality mitigation led to a partnership with the National Park Service and the North Cascades Institute to build the North Cascades Environmental Learning Center on the north shore of Diablo Lake.

The road on art-deco Diablo Dam leads to the North Cascades Environmental Learning Center.

The Gorge Dam powerhouse, where turbines from New York and generators from Pittsburgh produce electricity for Seattle.

Diablo Dam at night, in front of Davis Peak

Colonial Peak
(7,771 feet)

Paul Bunyan's
Stump (7,480 feet)

Pyramid Peak
(7,182 feet)

Davis Peak
(7,051 feet)

Thunder Lake Fault

Sourdough
Mountain
(6,120 feet)

Thunder Creek flows into Diablo Lake from the south, delivering ground-up "rock flour" from the Inspiration, Neve, Klawatti, and Boston Glaciers and half a dozen smaller glaciers. Thunder Creek follows the Thunder Lake Fault into Diablo Lake. The fault continues underneath the lake, emerging again as the slight divide bisecting Sourdough Mountain.

EXTREME HEIGHTS

- Snowfield Peak rises more than 7,000 feet above Diablo Lake.
- The northeast face of Davis Peak (named after 1890s homesteader Lucinda Davis) drops 5,250 feet in one horizontal mile, one of the longest vertical drops in the contiguous United States.
- The 5.5-mile hike up Sourdough Mountain has a vertical elevation gain of almost a mile (5,085 feet), comparable to hiking from the inner gorge to the rim of the Grand Canyon.

Across the lake, tucked into a cove on the north shore, is the North Cascades Environmental Learning Center. Here the North Cascades Institute provides seminars, retreats, and programs for individuals, families, and schools on natural and cultural history, art, science, and literature. Information at www.ncascades.org.

Rock-flour-tinted water is sometimes seen near the town of Darrington, where the Suiattle and Sauk Rivers run together. The Suiattle carries glacial silt from Glacier Peak.

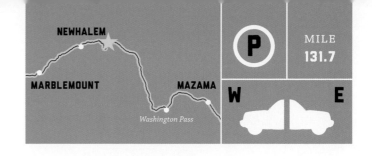

Diablo Lake LAKE ELEVATION: 1,200 FEET OVERLOOK ELEVATION: 1,720 FEET

A pallette of geologic processes produced this scene. The rocks now exposed in the mountains formed deep in the earth's crust, where rocks bake and minerals transform. Glaciers sculpt the highest peaks, producing rock flour, which colors the lake. An ancient fault slices the scene from Thunder Creek valley to Sourdough Mountain and beyond. The lake itself is new. Before Diablo Dam was finished in 1930, Thunder Creek joined the Skagit River here, which then flowed west to the Puget Sound.

A COLLOIDAL SITUATION

STARTLED by the stunning teal color of Diablo Lake, my artist wife mused on how nature doesn't always follow the rules we perceive as natural. Not only would this lake's tint look strange in a painting, but also what you see today may change next month. Or tomorrow morning.

Listening in the overlook parking lot, I've heard interesting explanations for the lake's color:

- Copper from nearby mines colors the water.
- Park Service officials put green dye in it.
- There's more photosynthesis in the summer.
- Because there's less oxygen in the mountains, algae grows.

The truth? It's related to marshmallows, milk, and mayonnaise. They, like Diablo Lake, are colloids—substances full of tiny suspended particles. The particles in this case are rock flour, ultrafine grains of rock ground under the mill of glacial ice high in alpine country. Carried as ultralight cargo by gravity, wind, water, and ice, the grains are delivered to the lake, where they hang out. Or hang in. The fine silt remains suspended in the water, causing Diablo Lake to be a grand-scale colloidal solution and shifting the color from blue to green.

In the late 1600s, Isaac Newton used prisms to show that "white" sunlight is really the combined colors of the rainbow. When sunlight strikes water, all of the colors are absorbed except those at the short-wavelength blue end of the light spectrum. Tiny particles in the water scatter this blue light back to the surface and our eyes. But silt particles cause a shift from blue to longer-wavelength green. The hue can change daily or seasonally because of the sun's angle and the amount of suspended sediment in the water. In late summer, after lots of glacial ice melt, the water is at its greenest. After autumn rains and cooling mountain temperatures, the lake clears up, becoming bluer again.

RECIPE FOR SKAGIT GNEISS

The surrounding peaks are made of what geologists call the Skagit Gneiss Complex, a group of rocks typical of the complex origins of the North Cascades. A former overlook sign humorously explains how to prepare Skagit Gneiss.

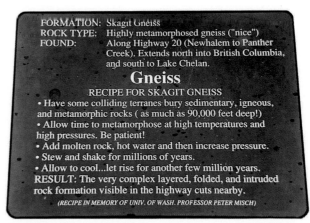

FORMATION: Skagit Gneiss
ROCK TYPE: Highly metamorphosed gneiss ("nice")
FOUND: Along Highway 20 (Newhalem to Panther Creek). Extends north into British Columbia, and south to Lake Chelan.

Gneiss
RECIPE FOR SKAGIT GNEISS
- Have some colliding terranes bury sedimentary, igneous, and metamorphic rocks (as much as 90,000 feet deep!)
- Allow time to metamorphose at high temperatures and high pressures. Be patient!
- Add molten rock, hot water and then increase pressure.
- Stew and shake for millions of years.
- Allow to cool...let rise for another few million years.
RESULT: The very complex layered, folded, and intruded rock formation visible in the highway cuts nearby.
(RECIPE IN MEMORY OF UNIV. OF WASH. PROFESSOR PETER MISCH.)

Colonial Peak
(7,771 feet)

Paul Bunyan's
Stump (7,480 feet)

Pyramid Peak
(7,182 feet)

Exposed face of Ruby Mountain

By reading the texture, composition, and structure of rocks, geologists decipher clues and look far back in time. Along the North Cascades Highway, you'll pass clues to colliding tectonic plates, ancient island chains, undersea volcanoes, and ice ages. In comparison, all the motorcycles, bicycles, cars, and trucks are as momentary as quarks. The mountains have their own time lines that stretch far into the past, and that will continue on long after our visit here.

Nearby Colonial Creek Campground has 162 sites hidden next to the calm lakeshore, exuberant streams, and majestic old-growth cedars.

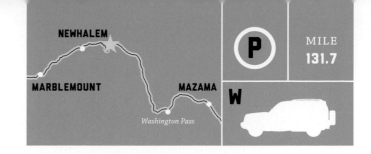
Colonial Peak 7,771 FEET
Pyramid Peak 7,182 FEET

Colonial and Pyramid Peaks stretch to heights where the sun's rays strike from dawn to dusk. Formed a hundred thousand feet below, this once-molten rock now scrapes the sky. Ice, water, and gravity unrelentingly attack the exposed faces, carving the jagged horns and knife-edge ridges that prompt our awe. These peaks were first climbed in 1931, decades before the highway or lightweight camping gear existed.

GEOLOGIC DETECTIVE STORIES

ALWAYS IMPRESSIVE, Colonial and Pyramid Peaks loom above the road. If we want mountain awe, they will do the job. They are formidable, solid, stable. But subtle clues reveal past turmoil in the tranquil scene: snow collecting on diagonal fractures formed from the tectonic stress of a rising range; triangular peaks and U-shaped depressions carved by a smothering cloak of glacial ice that slowly excavated the entombed mountains underneath; light patches of talus, temporary resting place for gravity-torn, cliff-born rocks from above. And far below, beside the road, a slice of ancient chaos is visible across from the parking lot, owing to a more recent (road building) disturbance: erosion by dynamite. The contorted innards of Ruby Mountain are exposed, revealing history in migmatite.

Geology is a detective story in time. When I look at a rock, I see ancient environments, shifting continents, subterranean chemical soups. Crystals in the rock across the road surrender tales of titanic forces. The dark rock has visible interlocking minerals, a clue revealing that great blobs of magma were once here, deep underground. The magma cooled slowly to become granodiorite, a rock similar to granite. The minerals are slightly aligned (foliated), indicating metamorphism to a rock called orthogneiss.

Another clue in this geologic narrative is the ribbony white bands that seem to flow within the dark rock. But how does one rock "flow" within another? The white stone is a mixture of quartz and feldspar, which both melt at temperatures of twelve hundred to fifteen hundred degrees Fahrenheit. Here, the rock was hydrothermally altered at much lower temperatures—partially melted while the surrounding rock was not. This cliff is made of hybrid rock, a combination of metamorphic and igneous. Regional pressure and heat deformed it all, twisting and kneading it like putty into contorted ribbons of migmatite, or "mixed rock." From granodiorite to orthogneiss to migmatitic orthogneiss.

INCOMPREHENSIBLE TIME

Geologic time and human time are hard to reconcile. The road has been here for decades. A motorcycle zooms by in seconds. The rock behind the motorcycle has been here seemingly forever—since long before our ancestors' ancestors. The ribbons of white in the rock are essentially unchanged after millions of years, but nothing is permanent in geologic time. The mountain is wearing away at a pace not noticed by human beings in a few trips across the mountains—or in the many lifetimes of generations.

Another incomprehensible time line exists inside the rock. Inside each individual grain. It is the microworld of quantum mechanics. Atoms vibrate. Electrons frenetically circuit nuclei. Quarks decay. All in billionths of trillionths of seconds. The rock, unchanged in human time, and virtually unchanged in tens of millions of years, is thrumming with trillions of interactions every second. All this activity takes place in rock that is still as stone.

Near Diablo Lake

Sourdough Mountain fire lookout tower surveys the scene at sunset. Ruby Mountain, too, catches the sun's last rays.

Arctic lupine (*Lupinus arcticus*)

August snowmelt keeps streams flowing.

A raven (*Corvus corax*) plays in the updrafts.

Late at night, Sirius, the "Dog Star," and Orion silently move past Colonial and Pyramid Peaks (seen from the overlook parking lot).

Alpine wildflowers on Sourdough Mountain

Sundew (*Drosera rotundifolia*)

Paul Bunyan's Stump in the early morning sun

Bracket fungi (*Fomitopsis pinicola*)

John Pierce Falls

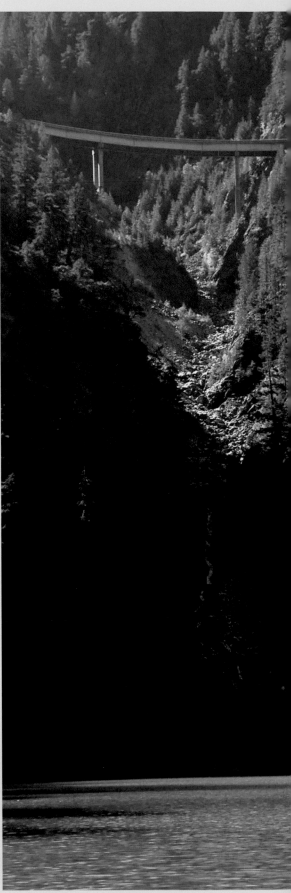

Highway 20 bridge at John Pierce Falls, viewed from
Diablo Lake, seven hundred feet below

John Pierce Falls

Horsetail Creek drips from snowmelt high on Ruby Mountain before plunging through this narrow chute, then under the road and over rocky talus to Diablo Lake seven hundred feet below. Along with Skagit Gorge, this stretch of road was the most difficult and dangerous to build. Construction equipment was buried under landslides and fell over cliffs. Workers died. The highway had to be completed, though, so cautious perseverence and ingenuity led to the road you drive today. The roadside cliff exposes complex metamporphic stone and stories of recycled atoms.

Everything from wet concrete to workers to 110-foot steel girders was lowered into place by cables.

EXTRAORDINARY NERVE led a foreman to drive his construction truck on a ledge so narrow that one of the back dual wheels hung over the cliff. That is how, in 1965, he emboldened his nervous construction crew to continue their work with heavy machinery at John Pierce Falls. In one incident, a vehicle's brakes failed and the driver jumped out before it fell into the abyss. Because this was such a hazardous and difficult chasm for road-builders to bridge, large cables and nets were rigged up to move beams, concrete, and even workers, who hung from boatswain's chairs. The waterfall is named after laborer John Pierce, who survived a dangerous fall.

Weather, avalanches, politics, geology, and economics forced eight decades of delays after the state first commissioned the road. The "North Cross State Highway" officially opened in 1972, the only cross-mountain access between Stevens Pass to the south and the Trans-Canada Highway (and two Customs stops) to the north.

ATOM STEW

This cliff face is a baked and scrambled puzzle of what used to be igneous rocks and ocean sediments. Squeezed, twisted, and contorted by the slow cookery of earth's internal heat and pressure, the rocks were transformed into banded gneiss, orthogneiss, paragneiss, amphibolite, migmatite, pegmatite, and biotite schist (included in the Skagit Gneiss Complex). They are all changed, all new forms of rock. But even with high temperatures and pressure that transform minerals and grains of stone, the gneisses and schists and pegmatites are all made up of recycled atoms. Along with hydrothermal additions, they are made from the same silicon, oxygen, aluminum, iron, and other atoms that were in the original granite, ocean floor, and ancient rocks of the earth's crust. They are composed of atoms that were forged in stars billions of years ago. Reused, recycled.

There is an exception. Radioactive uranium atoms decay into lead atoms at a specific rate. Geochronologists measure the relative amounts of uranium and lead to determine the ages of rocks, which at John Pierce Falls vary from forty-five to ninety million years. (For more information on radiometric dating, see appendix C.)

Recycled atoms in the Skagit Gneiss Complex

Hozomeen Mountain
(8,071 feet)

Jack Mountain
(9,066 feet)

The Skagit River used to meander across this valley with wide loops and oxbows. As Ross Dam (originally Ruby Dam) rose in the 1930s and 1940s, the valley flooded, including more than five thousand acres of Canadian land.

To the northwest stands 7,660-foot Mount Prophet, named by an 1890s gold miner and religious proselytizer, Tommy Roland. Names of nearby features include the Apostle, the Sinner, the Saint, Genesis Peak, Elijah Ridge, and Gabriel Peak.

Before Ross Dam was built, the upper Skagit River meandered across the valley.

Ross Lake LIFE

Glaciers receded from this valley thirteen thousand years ago, allowing life to return. Portraits of the past are buried in layers: glacial sediments preserve insect and plant fossils; archaeological digs reveal activities of human occupants; pollen in lake-bottom mud records vegetation and climate change. Cool, dry forests gave way to the warmer and wetter fir, hemlock, and cedar forests of today. Humans have visited and found resources here for at least eight thousand years.

SUPERMARKET AND TOOLBOX

LONG BEFORE THIS VALLEY WAS DAMMED, Skagit people journeyed along the meandering Skagit River. Where twenty-first-century hikers go to get away from it all was a highway system and shopping network for early inhabitants. They traversed these valleys and ridges to gather and process resources: plants, fish, birds, and mammals (lily root, huckleberry, trout, grouse, deer, bear, weasel, and much more). Some resources, such as chert and goat wool, were collected for trade with other Salishan-speaking peoples.

Archaeological digs in the area reveal five-thousand-year-old camps. Fire pits; cutting, crushing, and scraping tools; and a variety of bones all suggest active camps for meat, grease, and marrow processing and for dressing hides. Tiny amounts of organic residue still exist in rock fractures. Protein analysis reveals a who's who of woodland and alpine animals, from snowshoe hare to mountain sheep. Soapstone pipe fragments, nuts, and seeds have also been found.

Artifacts of another kind are found at higher elevations. More than two hundred million years ago, trillions of microscopic radiolarians died, sank, and left their silica-rich skeletons on the seafloor. Bulldozed by

tectonic orogeny, the seafloor became local hills containing chert—the microcrystalline quartz remains of radiolaria, and an excellent rock for flaking into sharp objects. Archaeologists found relatively recent to eight-thousand-year-old chert-processing camps high in these hills. Fire-pit charcoal is dated by measuring isotopes of carbon.

MODERN VISITORS

Seven thousand feet below Jack Mountain, Jack Rowley flipped rocks in chilly Ruby Creek while looking for nuggets of gold. A eureka moment led to success on the Original Discovery claim, which, along with the nearby Nip and Tuck mine, set off the 1880s gold rush.

More than a century later, canoers and kayakers float the lake, hikers walk the hills, and drivers stop and gaze from roadside overlooks—all perhaps looking for riches around Ross Lake.

Seventeen miles north stands Hozomeen Mountain, which entranced writer Jack Kerouac when he spent the summer of 1956 in a nearby fire tower on Desolation Peak. His time here led to two novels: *The Dharma Bums* and *Desolation Angels*.

Ross Lake sediments hide clues to past climate.

An upper Skagit River valley hearth more than five thousand years old, containing stone tools and deer bone and teeth fragments

A fresh layer of October snow adds dramatic contrast to 9,066-foot Jack Mountain. A thrust fault in the Ross Lake Fault Zone separates the upper mountain's former lava flows and ocean sediments from metamorphosed mica schist, which also was once sea-bottom sediment.

On Ross Lake's wet west side, Big Beaver Valley was eroded by glaciers into a broad U shape.

On the east side, narrow Devil's Creek was unaffected by glaciers.

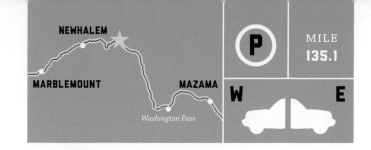

Ross Lake LAND

Climate sculpts what geology constructs. Ross Lake spans the Ross Lake Fault Zone, a geologic weakness surrounded by mountains composed of ancient ocean floor. Over long periods of time, rain and glaciers persistently carved the geologic landscape into ramparts that inspire climbers, photographers, and poets. Just west of here, high summits force moisture-laden Pacific air to unload. When the climate cooled, glaciers formed, severely eroding peaks and valleys; but eastward there was less snow, fewer glaciers, less erosion. Eventually the Skagit River cut meandering S-curves across the wide valley, which is now hidden under Ross Lake.

A COALITION OF GEOLOGY AND CLIMATE

Hozomeen's sharp peaks were nunataks above the glacial ice.

WELCOME to the Ross Lake Fault Zone, where rocks were sheared and broken, strained by tectonic stress, their weakness exploited by water and ice-age glaciers.

Evidence of the glaciers' past height can barely be seen, an eroded bathtub ring high above Ross Lake. All here was covered by ice except the lone high peaks known as nunataks. The sharp pinnacles of Mount Prophet, Hozomeen Mountain, and Jack Mountain rise above the former glacial ice. This boundary between rounded hills and jagged peaks marks the upper limit of the great Cordilleran ice sheet that flowed from Canada.

Ice didn't carve the west and east sides of this valley equally. Both sides were exposed to the same ice-age temperatures but not to the same precipitation. Again, geology is the fundamental reason. In this region, the mountains block wet marine air from eastern Washington. This effect is compressed in the Ross Lake basin. Just to the west are the Pickets, a sub-range in the North Cascades. They, volcanic Mount Baker, and other high peaks rise high enough to intercept rain and snow from Pacific storms, leaving little for the east side. During the Ice Age, this meant valley glaciers grew to significant size, carving out wide U-shaped valleys on the west side of what is now Ross Lake. The east side received less precipitation, forming less glacial ice, so the valleys were not cut as wide. To this day, the contrast on either side of the lake is striking. Not visible from the roadside turnouts, Big Beaver and Little Beaver Valleys' broad profiles form grand U-shaped entrances. On the other side of the lake, Devil's Creek and Three Fools Creek are narrow clefts in the mountainside.

HOW MUCH RAIN?

Even today, the Ross Lake area has one of the highest precipitation gradients in North America. Midway up the lake, Lightning Creek is in the rain shadow of the Picket Range. Receiving twenty-five to thirty inches of rain a year, it has dryland plants such as ponderosa pine and Rocky Mountain juniper, yet ten miles south, Ross Dam collects about sixty inches of rain, and ten miles farther west, Newhalem is drenched with eighty inches.

A small remnant of the Ice Age, Nohokomeen Glacier survives on the shadowy northeast side of Jack Mountain.

Ponderosa pine on the dry east side of the mountains.

The highway parallels the river, which obeys Newton's first law of motion. Simple forces relentlessly prevent the water from moving in a straight path.

Granite Creek Canyon

Mile after mile, the highway hugs the river, twisting and turning to match the riverbed, where the current bounces back and forth like a pinball between long canyon walls. But our senses tell us the opposite of what is true, for Granite Creek Canyon is straight. For about eight miles, the road parallels curvy Granite Creek, but road and river together strike northwest-southeast along what may be remnants of a prehistoric fault in the Ross Lake Fault Zone.

OBEYING NEWTON'S LAW

AS A YOUNG BOY, I sat in the middle of the back seat getting crushed by older brothers when the car turned sharp corners. Although unaware of it, we were obeying Newton's first law of motion: objects move in a straight line unless a force changes their motion. If not for seat belts, backseat siblings would have a field day on Granite Creek Canyon curves.

The curves themselves are a result of Newton's first law. Water in Granite Creek turns because it is relentlessly redirected by simple forces, whether from a fallen tree, a large rock, an entering tributary stream. The current deflects from side to side, eroding the riverbank. The bank drops rocks into the river, changing the river's course again, eroding the bank again, changing the river's course again. The dance continues day and night as the Cascades submit to perpetual surgery.

GOLD

Gold pan, Newhalem Visitor Center

Almost a hundred years before the road was built, prospectors explored these twisting canyons, looking for flecks of gold and pockets of fortune. In 1879 two claims, the Nip and Tuck and the Original Discovery, produced about twenty-five hundred dollars, starting a gold rush on Ruby Creek near what is now Mile 140 (9,066-foot Jack Mountain was named after one of the original miners, Jack Rowley). An onslaught of prospectors made hundreds of claims, digging, panning, and sluicing nearly a

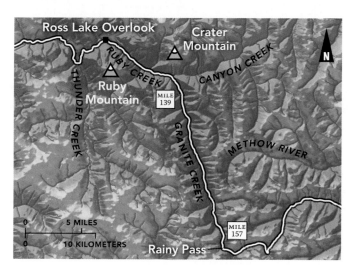

hundred thousand dollars (present day value) of loose gold out of the streams. But within two years, the easy gold was gone, and most of the prospectors left to seek their fortunes in British Columbia or Alaska. A second wave of miners came in the 1890s with heavy machinery to dig, blast, and drill into solid rock, looking for the sources of the stream gold. Once more, few prospects were profitable, because deposits were thin and it was difficult to get equipment and ore through the perilous Skagit River Gorge. Again, most of the miners left.

Not all came to get rich, though. In 1895 a Virginian came to stay. Forty-one years after being born into slavery, George Holmes found this wilderness and built a cabin. He leased the Original Discovery mine until just before he died, thirty years later. Coming out of the mountains to sell his gold to a Seattle jeweler every two to three years, he found peace in the hills and forests of Ruby Creek.

Jack Mountain
(9,066 feet)

Crater Mountain
(8,128 feet)

A former fire-lookout site, Crater Mountain looms six thousand feet above the highway. Photo taken at mile 147.7. A parking pullout is at mile 146.7.

A hiker looks down at Granite Creek Canyon and the North Cascades Highway from Crater Mountain.

Crater Mountain's "crater" is a glacially-carved cirque. This lake-filled depression lies four thousand feet above the road.

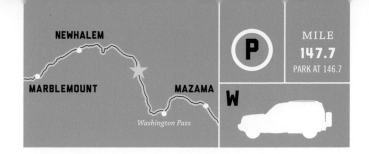

NEWHALEM

MARBLEMOUNT   MAZAMA

Washington Pass

P   MILE
    147.7
    PARK AT 146.7

W

Crater Mountain 8,128 FEET

Composed partially of volcanic rock and resembling a volcano that has met the ravages of time, Crater Mountain is, surprisingly, not a volcano. Situated next to the Ross Lake Fault Zone, Crater's rocks include former ocean lava and sediments that have been metamorphosed, faulted, and thrust to great heights. This peak had the highest operating fire lookout in the state, poet Gary Snyder's 1952 summer home. The six-thousand-foot climb to the top offers a stimulating rock scramble and stupendous views.

All that remains is the post that held the fire locating device.

MOUNTAIN SECLUSION and tranquility drew 1950s Beat Generation writers Gary Snyder, Phil Walen, and Jack Kerouac to Crater Mountain, Sourdough Mountain, and Desolation Peak. On these isolated mountaintops, they had time and solitude to study Zen Buddhism, write haiku, practice calligraphy, and meditate. Today, all that remains of Snyder's Crater Mountain cabin is the center post, part of the fire-sighting mechanism. The cabin on nearby Sourdough Mountain (manned by Walen and Snyder) remains, as does Kerouac's tiny cabin home far up Ross Lake on Desolation Peak.

Crater Mountain's fire lookout was one of the highest of more than six hundred built in Washington through the 1930s. It was actually too high to be of good use, because of the frequently cloudy weather, late snowmelt, and early snowfall (even in July). But it was perfect for Snyder's spiritual introspection. Reflecting on the difficulties of slowing down in urban life, he said the time he spent isolated on Crater

Sourdough Mountain lookout

Mountain "was the first opportunity I had to really see if I could sit."

OCEAN FLOOR NO MORE

An ancient window to undersea volcanism is exposed high on Crater Mountain's dry, steep slopes: pillow lava. Looking like a macabre stack of bowling balls, pillow basalt forms when molten lava flows into the cold sea and quickly hardens into round blobs.

Pillow lava on Crater Mountain

As in much of the North Cascades Range, Crater's rocks are transplants. Like Jack Mountain, Crater is split by a thrust fault. The top half is "greenstone," three-hundred-million-year-old metamorphosed basalt. Part of an ancient seafloor, the lava was transformed by heat and pressure when slowly shoved into the continent and pushed to where it now sits, eight thousand feet above sea level. The lower mountain is made from ocean-bottom mud compressed into shale, then slate, then phyllite, notable for its coarser-grained crystallized mica, which reflects sunlight.

Crater Mountain's composition of deformed oceanic rocks reminds us of the massive and extremely long-term forces creating these mountains. Made from materials of oceanic origin, tectonically displaced, which will eventually return to the sea, the North Cascades are as temporary as reflective summers spent in their midst—it's just a matter of one's timescale.

Crater Mountain Views

Jack Mountain
(9,066 feet)

Krummholz (stunted, windswept trees)

After-storm rainbow

Hozomeen Mountain

Bonanza Peak
(9,511 feet; thirty-five
miles away!)

Kimtah Peak
(8,620 feet)

Mount Goode
(9,220 feet)

Windbreak for a campsite with a view. There is no water on top of Crater Mountain.

The remains of a pulley system used to haul supplies up to the lookout tower

Glacial tarn (former eastern edge of the Jerry glacier)

Shade-tolerant Pacific silver fir (*Abies amabilis*) and western hemlock (*Tsuga heterophylla*) wait in the dark understory.

Black lines trace the high ridges seen on either side of Route 20 along Granite Creek. This view from a mountain named Golden Horn shows what is on the back side of the ridges, not seen from the road. The valley in the foreground of this picture contains the headwaters of the Methow River.

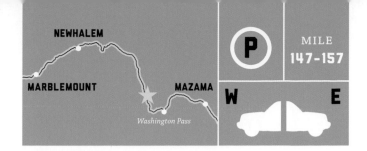

Upper Granite Creek

Fleeting breaks in the trees provide glimpses of a steep-walled southern horizon. I used to wonder: what's beyond those precipitous ridges? The cliffs seem close but insurmountable. What are they hiding on the other side? Waves. Thousands of acres of waves made of ridge after ridge of exposed granite, gneiss, and, to the north, sediments of the ancient Methow Ocean. Far below, at car-level elevation, your steady companion is miles of deep dark forest—closer than the rock faces above but almost as mysterious. Inert and idle-looking, these woods are massive carbon sinks, removing carbon dioxide from the atmosphere.

RECYCLED OXYGEN

THE FOREST CAME from thin air. Literally. Remember learning in school that trees take in carbon dioxide and give off oxygen? And that we do the opposite? We breathe in oxygen and exhale carbon dioxide. (Actually, most of what goes in and out is nitrogen.) But there's a missing piece of this story. If trees take in CO_2 and release O_2, what happens to the C? The carbon. Where does it go? Atoms can't just disappear. Through the marvel of photosynthesis, plants keep the carbon and release oxygen. The carbon *becomes* the tree. Oxygen is a tree's waste product.

The *mass* of a tree—the trunk, the leaves, the branches, the xylem and phloem, the heartwood and cambium—it comes from the air. Trees are a major carbon "sink," removing carbon dioxide from the atmosphere, impounding the carbon, and returning oxygen. How's that for recycling? There's so much biomass in the big trees here that more carbon is stored per acre in Pacific Northwest forests than in tropical rainforests. The national parks and national forests of Washington, Oregon, and Alaska are among the largest carbon banks on earth.

THE THOUSAND-YEAR WAR

A ceaseless battle is unhurriedly raging in the woods. There's no gnashing of teeth or torn fur or feathers. The trees are fighting for that precious resource, light.

Douglas fir grows quickly in the light. This trait, along with its thick, fire-resistant bark, makes it the dominant species blanketing Cascades hills. But waiting in the darkness are shade-tolerant Pacific silver fir and western hemlock. Patient they are. Doug firs don't regenerate well under the dark forest canopy, but hemlock can linger for tens or hundreds of years in the shadows, then have a growth spurt as light opens up when older trees fall over. Hemlocks suppress their own growth again when the canopy closes in from taller neighboring trees, repeating this cycle for hundreds of years. Hemlock seedlings litter the forest floor, covering nurse logs and growing in piles of bark fragments around the base of Douglas firs. *Eventually*, western hemlocks (and western red cedar) become the dominant trees in millennia-old climax forests. This doesn't mark the end, though, because in a forest's span of thousands of years, climate conditions can change, possibly favoring other trees. The leisurely combat continues.

Upper Granite Creek
Peaks along the Highway

Black Peak
(8,970 feet)

Repulse Peak
(7,923 feet)

Fisher Pe
(8,040 fe

Mile 154: A ring of peaks surrounds a glacially carved basin above the North Cascades Highway. There is a wide turnoff where you can stop to see this view at mile 153.9. Look slightly west to see Graybeard Peak.

Mile 154: Alpenglow on Fisher Peak

Miles 150–153: Porcupine Peak (7,762 feet) rises above the highway and thousands of acres of forest. A few miles up the road, the east face of Porcupine towers above Rainy Pass.

Miles 156–157: Distinctive Corteo Peak (8,080 feet) can be seen poking through the trees.

GRAY AND BLACK

Little-known 7,965-foot Graybeard Peak keeps poking up along a twelve-mile stretch of Route 20. Black Peak, at 8,970 feet, is the seventeenth-highest peak in Washington State. Blackbeard Peak sits lower and closer to the highway.

Mile 146: The northwest face of Graybeard Peak, seven miles away

Mile 154: The northeast face of Graybeard. People climb and have even skied this peak.

Black Peak rises from Heather Pass, a 2.3-mile, moderately steep hike from the highway. There is still plenty of snow on this August day.

Black Peak (8,970 feet)   Blackbeard Peak (7,241 feet)   Graybeard Peak (7,965 feet)

Mile 157: Black Peak, Blackbeard Peak, Graybeard Peak

Graybeard Peak   Fisher Peak

Looking east at Graybeard and 8,040-foot Fisher Peak from Easy Pass

Forest and Meadow Treasure

Hackelia (Jessica sticktight)
(*Hackelia micrantha*)

Western meadowrue (female flowers)
(*Thalictrum occidentale*)

Prince's-pine pipsissewa
(*Chimaphila umbellata*)

Bracted lousewort (*Pedicularis bracteosa*)

Foam flower (*Tiarella trifoliata*)

False hellebore (*Veratrum viride*)

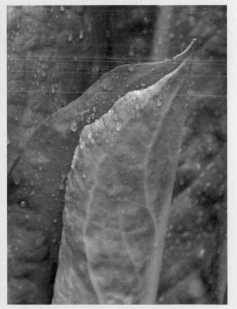

White-veined wintergreen (*Pyrola picta*)

Arctic lupine (*Lupinus arcticus*)

Skunk cabbage (*Lysichiton americanum*)

45

The east face of 7,762-foot Porcupine Peak towers above the road.

West meets east at Rainy Pass. A blanket of condensed moisture keeps the west side in cool clouds, while it's dry and warm just to the east. Campers shivering high above Rainy Pass on this day didn't know the sun had been shining all morning fairly close by.

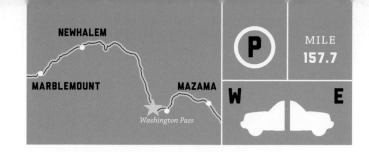

MARBLEMOUNT MAZAMA

Washington Pass

P MILE 157.7

W E

Rainy Pass 4,855 FEET

Rainy Pass is the final barrier to moisture-filled clouds from the Pacific, which drop a yearly average of fifty-six inches of rain. This pass separates the wet west from the dry east. From here, the road follows water downhill, east to arid scrublands or west to dense green forests. Far away from the road, elusive animals live at high elevation, surviving harsh winters. Seen by few people, ptarmigan and pika both rely on cool temperatures for their survival. They are indicator species for the health of alpine ecosystems.

FROM THE SEA AND BACK

I SAT ENTRANCED one morning watching a steady flow of clouds move over the pass and disappear before my eyes. It was a cloud finish line. This is where tides of moisture-filled air rise from the Pacific Ocean and lap the shore of the Cascades crest. Crossing this divide, the air sinks and warms. Moisture evaporates, clouds dissipate, and the air continues east, sinking to lower elevation and higher atmospheric pressure.

Rain and snowmelt at Rainy Pass make two distinct journeys to the ocean. Flowing east, water streams past lava beds and dry land, irrigates orchards, joins the mighty Columbia River, passes over nine major dams, and finally pours into the Pacific Ocean more than five hundred river miles away. To the west, water passes through green moss-draped forests and the fertile Skagit Valley on the way to Puget Sound, a little more than a hundred river miles away. Like the salmon they sustain, countless water molecules spend time in the Pacific before returning to Skagit headwaters.

ESCAPING THE HEAT

If you're a pika, your survival depends on cool temperatures. Home is high-elevation talus patches with cool microclimates under the rocks. Pikas have to regularly forage for food but, because of high internal body temperature and thick

fur, must remain below the surface on hot summer days. Recent research of more than a hundred talus patches in the national park found the greatest populations when temperature minimums beneath the talus were below about fifty-five degrees Fahrenheit. Because of their restricted

Scientists set up a remote data logger to record temperature 144 times a day, 365 days a year.

range and sensitivity to temperature, snow line, and plant abundance, pikas are considered a climate-change indicator species. Those living at low elevations and on south aspects appear to be the most vulnerable.

Another climate-sensitive resident is the white-tailed ptarmigan. Exclusive to North America these birds live in high, rocky alpine areas and are seldom seen because of their remote habitat and camouflage.

Although their secluded environment keeps populations fairly safe from human impact, warmer temperatures caused by large-scale climate change pose a significant threat to ptarmigan populations. Breeding capability, metabolic stability, and habitat (in particular, above-treeline soft snow for roosting) are at risk.

A pika gathers vegetation to store for winter food.

A white-tailed ptarmigan molting its white winter feathers.

47

Rainy Pass Area Trails

Rainy Pass is a trail hub for hikers of all levels. Begin the shortest or longest of walks here—a paved stroll or a trek all the way to Mexico. From Rainy Pass, the Pacific Crest Trail leads north and south along the backbone of the most rugged part of the Cascades range.

EASIEST

Most gentle is the 0.9-mile paved, wheelchair-accessible trail leading to Rainy Lake. Interpretive signs describe unique characteristics of mountain vegetation. It doesn't get much easier to hike, picnic, and listen to waterfalls around a cliff-lined lake.

Benches and interpretive signs line the paved trail to Rainy Lake.

MOST DIFFICULT

A stone's throw away is the 2,650-mile Pacific Crest Trail, connecting Mexico and Canada. Not for the faint of heart, it has greater elevation changes than any other National Scenic Trail in America. Close to two hundred hikers each year take the five to six months necessary to complete the full journey. Expect to see through-hikers crossing Rainy Pass in September or October.

IN BETWEEN

Not up for a paved path or multistate expedition? Various day hikes begin on both sides of the road.

Alive with wildflowers in August, the two-mile hike to sparkling Lake Ann starts at the Rainy Lake trailhead. A 7.2-mile loop trail continues to Heather Pass, Maple Pass, and back again to the parking lot. A far more rigorous and technical challenge is 8,970-foot Black Peak. From Heather Pass, the three-mile approach features a hazardous boulder field, cliffs, snowfields, two lakes, and, depending on the route, a final two-thousand-foot climb.

A hiker enjoys wildflowers on the Heather Pass trail.

After starting near Mexico in April, a through-hiker pauses in October, six months and 2,580 miles into his journey on the Pacific Crest Trail.

Harsh paintbrush (*Castilleja hispida*)

Glacier lily (*Erythronium grandiflorum*)

Red columbine (*Aquilegia formosa*)

Black Peak rises above Heather Pass.

On the north side of the road, a five-mile hike on the Pacific Crest Trail leads to 6,800-foot Cutthroat Pass. Once there, roam ridges of yellow Golden Horn granite and take in the waves of peaks in every direction. Hikes can extend for days into the Pasayten Wilderness or all the way to Canada, the final destination for those hiking the trail's full length.

Cutthroat Pass hikers gaze at Golden Horn (8,366 feet) and Tower Mountain (8,444 feet).

The Milky Way glows above Cutthroat Pass.

The highway lies silently under the clouds as the sun rises on Porcupine Peak, Cutthroat Pass. (Dome and Corteo Peaks are in the background.)

Hiking isn't the only way to travel by foot. Backcountry skiers and snowshoers take advantage of heavy snow in the region, exploring valleys and peaks into July. Since the highway is typically closed from November through April (or later), most winter users wait for the road to open. Expertise in route-finding and avalanche safety is essential.

Skiers climb up, then ski down from Heather Pass on July 4.

49

Rainy Pass Views

Rainy Peak (7,768 feet)

Frisco Mountain (7,720 feet)

Early morning at Rainy Lake, a 0.9-mile paved walk from the parking lot.

Not visible from Rainy Lake, the Lyall Glacier once clung to the flanks of Rainy Peak. Less than three miles away, the Lewis Glacier filled a basin near Black Peak. The last vestiges of both melted away in the 1990s and early 2000s, casualties of a warming climate. Melting ice from these glaciers no longer supplies water to streams and lakes during the summer dry season (including the large waterfall flowing into Rainy Lake).

Rainy Lake waterfall

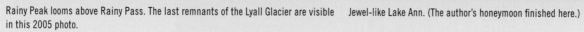

A hiker looks back at Rainy Pass. The North Cascades Highway arcs past the low end of this valley.

Rainy Peak looms above Rainy Pass. The last remnants of the Lyall Glacier are visible in this 2005 photo.

Jewel-like Lake Ann. (The author's honeymoon finished here.)

McGregor Mountain under a sky of ice-crystal-loaded cirrus clouds, harbingers of an approaching warm front.

Switchblade Peak	Stiletto Peak	Fire tower site
(7,805 feet)	(7,660 feet)	(7,223 feet)

Stiletto Peak's sharp ridgeline hides Jackknife Peak and Dagger Lake.

Stiletto Peak 7.660 FEET
McGregor Mountain 8.122 FEET

Walk down the valley, wind around McGregor Mountain, and twenty-two miles later, saunter into the remote village of Stehekin, by Lake Chelan. In the Interior Salish, or Nxaámxcín, language, Stehekin *(stxʷíkn')* means "the way through." The Stehekin valley was a corridor from the dry eastern interior to the high peaks and wet west side of the mountains. These valleys were *the way through*. A 1933 survey selected a route here for the future road, linking Washington Pass, the Upper Stehekin Valley, and Cascade Pass to what is now Marblemount. This area is also a juncture of public lands, managed by four federal agencies with different purposes.

WHO'S IN CHARGE?

WEAVING ACROSS these valleys and ridges is a confluence of national park, national forest, wilderness, and national recreation areas. Face south and scan the horizon—you're looking at four different types of public land. The north side of Stiletto Peak (and the road) is in Okanogan National Forest. The south side is in the eastern tip of North Cascades National Park. Stiletto's east edge lies on the fringe of the Lake Chelan–Sawtooth Wilderness. McGregor Mountain spans North Cascades National Park and the Lake Chelan National Recreation Area. These areas are managed by different branches of government and have different rules that allow for different activities. Wilderness areas alone are managed by four different agencies: the National Park Service, the U.S. Forest Service, the U.S. Fish and Wildlife Service, and the Bureau of Land Management.

Even along the road, it's not always clear. Thought you were driving through North Cascades National Park? You were close, but never really in it. Because of preexisting dams and the villages of Newhalem and Diablo, the official park boundary is on either side of the road and surrounds the villages. Along one stretch, you pass through the Ross Lake National Recreation Area, and currently you are in the Okanogan National Forest.

STILETTO PEAK

Stiletto is neighbor to the appropriately named Jackknife Peak (7,700 feet) and Switchblade Peak (the highest of the three at 7,805 feet). Dagger Lake sits fifteen hundred feet below Stiletto. A lookout with 360-degree views once perched upon the ridge. Access is difficult, because an old, steep, unmaintained trail disappears partway up the mountain.

McGREGOR MOUNTAIN

Named after Billy McGregor, an 1890s Stehekin Valley resident, it, too, had a summit lookout in the 1920s. More than a hundred steep, dry switchbacks climb sixty-three hundred feet to loose rock near the top. McGregor is mostly made from the same Skagit Gneiss Complex rocks as the Diablo Lake area twenty-five miles northwest. This geologic unit parallels the highway, separated from the road by much younger intrusions of Golden Horn granite.

Fire lookout on McGregor Mountain

Streamlined firs and shedding larches, each ready in its own way for impending deep winter snow

A Golden Horn granite dike cuts across Black Peak gran-ite on next-door Whistler Mountain. Golden Horn granite is forty-five million years younger than its darker host.

Stars slowly move across the northern sky.

Cutthroat Peak 8,050 FEET

The wide turnout and expansive view make this a popular roadside stop. First summited in 1937, this steep granite peak has climbing routes on all sides but no official trails. Seasons clearly progress as the basin melts in late spring, followed by tapestries of summer green, spectacular August wildflowers, and autumn's reds and golds. Snowfalls create winter's tricolored canvas of white, dark rock, and azure sky.

PART OF THE VIEW

IT WAS THE VASTNESS THAT GOT ME. I became part of the scene instead of an observer. It's an impressive view from the road, no doubt, and one that I remember from the honeymoon my wife and I took so many years ago. A *Sound of Music* view. Looking up, seeing the grass-covered meadow and high, ragged peaks, it's easy to imagine yourself up there, arms spread, feeling alive with the hills. For most, the reality is: love the view, take pictures, inhale the mountain air, and move on.

Being *part* of the view though . . . That means committing to the steep walk down through the trees, rock-hopping the cold stream, and putting quadricep muscles in burn mode while zigzagging up the hillside on a steep climbers' trail. Marmots and marmot holes dot the hillside. Their whistles warn. Grasses give way to piles of granite shreds from a mountain falling apart.

I sit on a prominent outcrop. There's space up here.

Dark, oxidized cliffs loom above and miniaturized cars silently glide below. The kaleidoscope of meadow flowers is gone, pollinators have flown away, seeds have dispersed, and dried husks are ready to recycle back into the thin soil. Slim, green firs cluster in front of me; golden larches watch my back. The firs up here are diet-thin, streamlined to shed winter's snow; larches prepare by stripping bare. Deciduous needles will soon carpet the ground.

Winter will muffle it all in deep white. Marmots in dark tunnels will hibernate on stored fat. Pikas under rock piles will live on their stashes of chewed-off plants. In another month the road will close and trillions of exquisite six-sided crystals (no two alike) will descend. For now, I'm content to sit in this cirque and soak up the mild autumn sun, being a dot in the scene before descending to my little toy car, where I'll watch flickering fifty-miles-per-hour views through my window.

August wildflowers—left and right, false hellebore (*Veratrum viride*)—grow below Cutthroat Peak. Across the valley, Early Winters Spires and Blue Lake basin rise above Highway 20.

North Early
Winters Spire
(7,760 feet)

Concord Tower
(7,560 feet)

South Early
Winters Spire
(7,807 feet)

Lexington Tower
(7,560 feet)

Liberty Bell
Mountain
(7,720 feet)

A two-minute walk to the fenced viewpoint seven hundred feet above the road reveals why the Park Service placed an overlook here. This late summer view captures the hairpin turn, Early Winters Spires, and Liberty Bell Mountain.

Washington
Pass
ELEVATION 5477

More than a mile high, Washington Pass is one of the highest paved roads in Washington State.

Mountain goats (*Oreamnos americanus*) live in precarious places, this one on the edge of Liberty Bell Mountain.

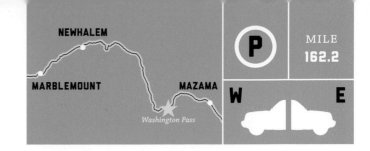

NEWHALEM
MARBLEMOUNT
MAZAMA
Washington Pass

P
W   E
MILE
162.2

Washington Pass 5,477 FEET

The most spectacular and most photographed slice of the North Cascades is dominated by the Liberty Bell and Early Winters Spires massif. Easy access, stunning views, dry climate, and ideal rock type make this area a favored destination for rock climbers. The steep walls here are a result of weaknesses in the rock and scouring by ice. Just a few thousand years ago (a blink in geologic time) glaciers filled these valleys, carving out the distinctive U shapes.

CLIMBERS PARADISE

EARTH'S EXOSKELETON rings the road here; diminutive humans climb every face. More than fifty published climbing routes exist just on the five peaks of the Early Winters Spires and Liberty Bell massif, most with plainly descriptive names such as East Buttress or Southwest Couloir. Others are more inspired: Tooth and Claw, Overexposure, Marginal Mystery Tour, Labor Pains. And some echo the Revolutionary War spirit of Liberty Bell Mountain and Lexington and Concord Towers: Freedom Rider, Independence, Midnight Ride.

Climbers scaled these peaks decades before the highway was here, with first ascents in 1937 (South Early Winters Spire) and 1946 (Liberty Bell). Numerous climbs start on the back (south) side but some of the more challenging routes face the highway. Look closely with binoculars and you might see tiny specks creeping to the top.

Much of the rock at Washington Pass is granite, which contains crystals of quartz and feldspar. These hard minerals have different shapes, which create a rough surface that is easy to grip. The granite here formed from cooling magma almost fifty million years ago. During uplift caused by tectonic forces, faults formed, mostly visible as the large breaks separating peaks. At the same time, overlying pressure was reduced, owing to uplift, causing cracks to form throughout the rock. The cracks are enlarged over time when water seeps in, freezes, and expands (one of many ways water wears down mountains). This bounty of faults and cracks provides a variety of routes to challenge climbers of all levels.

Even for nonclimbers, the steep ramparts, transient alpine lighting, and abundant viewpoints are raw ingredients for awe and inspiration. It's easy. Take a short walk, or just pause and savor the scene. What better place to breathe mountain air?

"Ice season," as climbers call it, is limited in Washington. Excellent conditions were found on a sunny November morning below Early Winters Spires.

The author enjoying a relaxing dinner at Washington Pass

Washington Pass Peaks

Scenic overlook

Glacier-carved U-shaped valley

Big Kangaroo (8,280 feet)

Wallabee Peak (7,995 feet)

Kangaroo Pass (6,671 feet)

View from dirt turnout near the top of the pass (at ○ on panoramic picture below), looking from north to east

Early Winters Spires and Liberty Bell Mountain

Scenic overlook

Tower Mountain (8,444 feet)

Golden Horn (8,366 feet)

The Needles (8,160 feet)

Big Kanga (8,280 fe

View from the top of Wallabee Peak, looking down on Washington Pass from west to north. The panoramic picture at the top of this page was taken from a location marked by the yellow dot. The photo at the top of the facing page was taken from a location marked by the green dot.

Vasiliki Tower
(7,920 feet)

Burgundy and
Chianti Spires
(each 8,400 feet)

Pernod and Chablis Spires
(8,441 feet and 8,350 feet)

West Peak,
Silver Star Mountain
(8,840 feet)

Silver Star Mountain
(8,876 fcet)

View from next to Washington Pass Overlook sign (at ◉ on panoramic picture below), looking northeast.

Half Moon
(7,960 feet)

West Peak,
Silver Star Mountain

Silver Star
Mountain

Smoke from Tripod Fire,
August 2006

Mt. Gardiner
(8,956 feet)

Washington Pass

South and North
Early Winters Spires

Liberty Bell
Mountain

Driving west from the Methow Valley. This spectacular approach to Washington Pass demonstrates why the region is called the American Alps.

Golden larches (*Larix occidentalis*) shine
in front of a cliff face below the Early
Winters Spires

South and North Early Winters Spires
(7,807 feet, 7,760 feet)

Lexington Tower
(7,560 feet)

Concord Tower
(7,560 feet)

Liberty Bell Mountain
(7,720 feet)

Early May view of the Early Winters Spires and Liberty Bell Mountain from nearby Silver Star Mountain

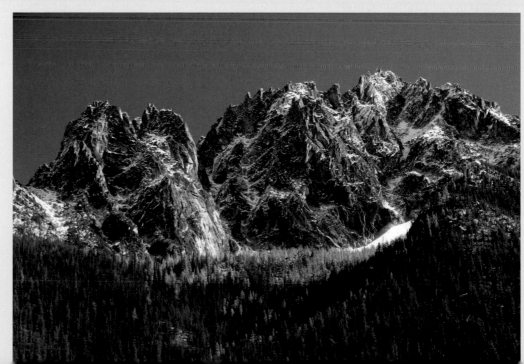

October larch trees (*Larix occidentalis*) in full glory beneath the Wine Spires and Silver Star Mountain

Washington Pass

The North Cascades Highway far below the scenic overlook. The Wine Spires and Silver Star Mountain stand in the background.

Glaciers carved this valley in the shape of a catenary curve, an elegant mathematical result of balanced physical forces. The Saint Louis Arch is a modified catenary curve, as are some modern tent designs.

Cutthroat Ridge
(6,978 feet)

Cutthroat Peak
(8,050 feet)

The Eastern Approach

Catch glimpses of high peaks through openings in the roadside walls of fir stands. Extraordinary sights are just around the next corner. And the next. And the next. For another fifty miles. The road soon ascends thousands of feet to Washington Pass and "America's Alps."

THE ROAD almost didn't go this way. In the 1890s, the highway was supposed to meander south of here, from the town of Twisp, over Twisp Pass, over Cascade Pass, and out the Cascade River valley to the town of Marblemount. The Cascade Pass plan, favored through the 1920s, eventually proved too steep and avalanche prone. Northern routes up Slate Pass mining trails were also considered and eliminated. The 1896 Board of State Road Commissioners' report describes "a slow hard trail, replete with rock canyons and high side hill work, while the grades on the old prospectors' trail were so heavy and numerous that, for the greater part of the way they would be of no use in constructing a road." It wasn't until the 1950s that the commisioners finally agreed on the Rainy Pass–Washington Pass route.

In winter, the highway closes at two points: nearby, at milepost 170.6, and near Ross Lake, at milepost 134. Deep snow and avalanches from peaks and ridges in the Washington Pass area usually shut down the highway from late November to late April. The earliest closing date was October 17, in 2003. The latest opening date was June 14, in 1974. A drought year in 1976–77 was the only year the highway remained open all winter. In the spring, the Department of Transportation's road-clearing crews

Kangaroo Ridge (8,182 feet)   Cutthroat Peak (8,050 feet)

bring in bulldozers, snowblowers, excavators, front-end loaders, and a World War II–era howitzer. To make areas safe for clearing, the howitzer shoots artillery rounds into unstable slopes, creating intentional avalanches. When areas are stable, crews bring in the heavy equipment and clear the road. Slides from Cutthroat Ridge, Liberty Bell Mountain, and Whistler Mountain can be especially deep and hazardous. Sometimes a pile or two of snow remains on the side of the road all summer, providing snowball fodder for hot August days.

A World War II howitzer used to create avalanches

One May day at Washington Pass, two weeks after the road opened

A vertical mile above the road, the Needles are a first glimpse of fifty miles of mountainous terrain.

The view from Granite Pass helps explain why these peaks are called the Needles. This light-colored rock is forty-eight-million-year-old Golden Horn granite.

The Needles 8,160 FEET

About forty-eight million years ago, molten rock with unusually high amounts of rare elements such as yttrium and cerium cooled and crystallized into Golden Horn granite. This yellow-tinted rock forms the cores of many peaks and ridges in the Washington Pass area, including the Early Winters Spires, Cutthroat Peak, and the Needles. Slowly eroding, their outlines are sharpened while water carries away the rock fragments. The river and highway both transport passengers from the high peaks to the dry country of eastern Washington.

THE PLACE IN BETWEEN

IT'S THE APPROACH . . . the transition . . . This stretch of road is the portage between North Cascades wilderness and a string of arid towns along the Methow River, dots leading to the vast open landscape of eastern Washington. Neither high peak nor lowland valley, it's a place of anticipation or relief, depending on the traveler's direction and taste for alpine heights.

The dry Methow Valley is, geologically, the extroverted cousin of the rain-soaked Skagit Valley west of the mountains. Like two reverse images in film, they have similar but converse appearances. Through both, the road and river meander together, leaking out of the mountains. But while the west side is hidden in green, the east is a sidewalk sale of geology, with goods easily visible for all who pause to look.

A CHANGE IN GEOLOGY

This stretch of highway links distinctly different geologic terranes. To the west are twisted, fractured, and scarred remains of tectonic collisions: gneisses, schists, and granitoids which result from high temperatures, pressures, and other relentless forces of tectonic activity. From Washington Pass to this low point, Golden Horn granite predominates. Erosion has reduced the mountain massif, washing sediments down the valley to be recycled into future sedimentary rocks.

East of mile 171 are layered sandstones, shales, and conglomerates, the geologic refuse carried by rivers draining an ancient pre–North America. This river-plain environment in the Methow Valley preserved something that is absent from the core of the North Cascades: fossils. West of here, extreme heat and pressure in igneous and metamorphic rock destroyed traces of ancient life, but conditions for retaining such traces were perfect along the broad coastal plains expanding and evolving here more than a hundred million years ago. Sediment from meandering rivers buried leaves, stems, and other detritus from ginkos, cycads, ferns—in all, more than 140 species of plants. Stratified, buried, and lithified, these fossils are now exposed in the cliffs and back hills of the Methow Valley.

West of here, igneous rock includes forty-eight-million-year old Golden Horn granite with vugs of crystals.

East of here, sedimentary rock includes sandstone with 105-million-year old fern fossils from the Rendezvous Mountain area.

| Driveway Butte (5,982 feet) | Last Chance Butte (7,046 feet) | Robinson Mountain (8,726 feet) | Goat Wall (approximately 4,500 feet at this point) |

Memorable North Cascades ventures start with a Mazama sunrise.

The Methow Valley Sport Trails Association sponsors biking, skiing, and running events throughout the year.

NEWHALEM

MARBLEMOUNT

MAZAMA

Washington Pass

MILE
179.6

Mazama Junction

It is sixty miles to the only traffic light in the county. There are glaciers closer than that.—Geof Child

Mazama is the last (or first) gas and food stop for sixty miles. Mountains visible here are an overture to a symphony of peaks and surprises—Driveway, Sandy, and Last Chance Buttes hide loftier and craggier mountains beyond. The valley transitioned from centuries of native occupation to an influx of prospectors and settlers in the 1880s. Mining came and went. Now orchards grow in the lower valley, and the upper valley is known for a world-class ski trail system and festivals featuring, among other things, hot air balloons and blues concerts.

THIS MARKS ONE END of the North Cascades journey. If you have a day to wander (and the nerves to negotiate Dead-Horse Point), turn here to Mazama, then go west over twenty-seven miles of mostly dirt road to Harts Pass and 7,400-foot Slate Peak, areas rich in history, geology, and adventure.

GOAT WALL, GREEN ROCK, AND GOLD MINES

The huge monolith of Goat Wall rises almost two thousand feet above the town of Mazama and parallels the road for close to five miles. Scraped and sheared by glaciers, the cliffs include some climbing routes, but the rock is crumblier than climbers' preferred granite. Close inspection reveals a metamorphosed type of lava called greenstone breccia, formed when minerals, altered by heat, recrystalized into green chlorite and epidote. Physical forces fractured the rock into angular fragments (breccia).

Injections of geothermally heated water deposited veins of quartz, sulfur, and metals in these hills. Metals such as gold. In the early 1930s, the Mazama Queen gold mine operated four hundred feet up the face of Goat Wall. Short-lived, it was one of many local mining operations attempted between the 1890s and the mid-1900s. Other local mines included the Mazama Pride, the American Flag, and the

Gold Key. At the foot of nearby Last Chance Butte stood a post office and the Last Chance Saloon, both of which served prospectors at the turn of the twentieth century in what was then the community of Robinson.

THE METHOW

The Methow Valley has been inhabited for almost ten thousand years. Distinct troughs at eighteen sites are evidence of early residents' pit houses. Trade with valley outsiders eventually brought in horses, canvas, and other useful items, but also smallpox, which, as it did elsewhere around the West, painfully reduced the native population. In 1880, presidential orders reserved this region for Chief Moses and local tribes, but six years later, the valley was officially opened to nonnative prospectors, trappers, loggers, and settlers. The Methow and other nearby tribes were moved from their ancestral lands to the Colville Reservation.

One hundred twenty-five years later, residents face conflicts between those who would develop and those who would protect the valley's natural beauty and open spaces. A rural quality of life with fresh air, freshwater, open spaces, and a lack of traffic have long been the character of this valley. Agriculture remains, but in the upper valley, outdoor recreation has become a key part of the economy, the quality of life, and the valley's sense of community. A new economic base, largely developed by valley residents, includes cross-country skiing, mountain biking, hiking, and numerous outdoor events and festivals.

FOUR VEINS CUT IN MAZAMA MINE

A face of good ore has been exposed in the Mazama Queen mine of the Continental Gold and Silver Mining company, 14 miles north of Winthrop, Wash., according to Ulysses Widman, president, Rosalia, and E. R. Krause, secretary-treasurer, Spokane, last week after an inspection.

Of $34.65 obtained from a sample, $30.80 was in gold, and the remainder in silver and zinc, according to a certificate they submitted. The face is five feet wide, they reported. When sampled, the face was at a point in the tunnel 533 feet from the surface

News clipping, 1936, from the *Spokesman-Review*, Spokane

Glaciers: Mountain Architects

ICE ON THE ROCKS

HORNS, ARÊTES, CIRQUES, hanging valleys. You've seen them high above the road: pyramid pinnacles, jagged ridges, and bowl-shaped depressions. Glaciers' remarkable ability to carve rock steepens and sharpens the landscape. But glacier ice isn't harder than granite. Water seeps into rock, freezes, and expands (like a soda can in the freezer), breaking the rock apart. The rock is then scraped up, or it falls onto the ice. As glaciers slowly flow downhill, ice and captive rocks scour valley walls, plucking more rock into the icy casserole.

LEFTOVERS

Glaciers are part excavator and part bulldozer. After plucking and scooping up rock, the ice carries and then leaves behind huge piles of jumbled rock called moraines. *Lateral moraines* are rock pushed to the valley sides. *Terminal moraines* mark the farthest point a glacier has traveled down-valley—where the

Lateral moraines. A dwindling glacier near Mount Redoubt left behind two-hundred-fifty-foot-high moraines.

conveyor belt dropped its captive stones. As glaciers in the North Cascades melt, moraines are left behind, signs of the ice's former extent.

U-SHAPED VALLEYS AND CATENARY CURVES

A typical river valley has a V-shaped profile; water flows in the bottom of the V. But glaciers carve downward and sideways, transforming the V into a steep U shape such as at Washington Pass. Not quite parabolic, the U shape is a

U-shaped valley and catenary curve. Looking down-valley from Washington Pass, at mile 162.5

catenary curve, an elegant mathematical result of balancing physical forces—in this case, maximum friction between ice and rock. A string or chain held between two hands may fall into a catenary curve. Some arches and modern tent designs are catenary curves, built for stability.

Glacial features on Silver Star Mountain, in the Washington Pass area

Walking across a glacier is a humbling and spirit-soaring experience. Ice axes, ropes, crampons, and the executive functions of our prefrontal cortex are protection against slipping into the abyss.

ICE SCULPTURES

Directions for creating glacial features you may see along the North Cascades Highway:

HORN: Sharply carve a peak on all sides until there's nothing left but a precipitous pinnacle. Favorite haunt of climbers.

ARÊTE: Carve a ridge on both sides until there's nothing left but a steep, serrated barrier. Give it a name like Ragged Ridge, Ripsaw Ridge, or Backbone. Arêtes often separate two cirques.

CIRQUE: As if you were a giant worm facing uphill, chew out the head of a valley, then digest and move the rock down-valley. The cirque is the chomped out, steep-sided bowl left behind after a glacier has feasted on rock.

TARN: Carve a depression in bedrock, usually in a cirque. Melt snow, add rain, let sit. Go for a swim. Brain freeze. (Tarns are usually at high elevation, so most are not visible from the road; hence they have names like Hidden Lake.)

HANGING VALLEY: Run a glacier from a high valley into a larger glacier in a lower valley. Let each glacier carve its valley, then warm the climate and melt the ice. The upper valley will be left "hanging" above the lower valley. Waterfalls often occur at hanging valleys, such as Bridalveil Falls in Yosemite National Park and many falls in the North Cascades.

Blue Lake (tarn), a two-mile hike from Washington Pass

Snagtooth and Kangaroo Ridges, near mile 167.2

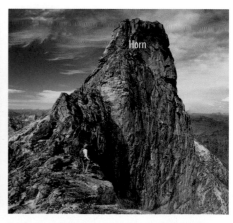

Golden Horn, 8,366 feet and an eleven-mile hike from Rainy Pass

The Coleman Glacier flows down the northwest side of Mount Baker.

A National Park glaciologist uses a World War II tank antenna to measure snow depth at 8,100 feet on Mount Spickard's Silver Glacier.

WALKING ON ICE

Geologists estimate that in the last 150 years about 40 percent of North Cascades National Park's ice cover has been lost. One study reports an ice volume loss of 20 to 40 percent in just the last twenty-five years. Fourteen North Cascades glaciers of a variety of sizes, elevations, and locations are measured multiple times a year. The rest are photographed from the air. Snow depth and ice volume are calculated using GPS, World War II tank antennas, and laser ranger finders, which determine distance to an object using a laser beam. Glaciers grow or shrink depending on yearly precipitation amounts and seasonal temperatures. Colder temperatures usually mean more snow, warmer more rain. Each glacier responds differently depending on elevation, slope, angle to the sun, and exposure to Pacific storms. Glaciers may grow or shrink in any given year, but the overall trend is that every North Cascades glacier is shrinking. The North Cascades data become part of worldwide glacier statistics, which help scientists understand global climate change. Locally, we gain a better understanding of climate impacts on glaciers and how North Cascades glaciers affect stream flows and ecosystems.

Glaciers Sustain Life

LIKE MOSS GROWING on the shady side of a tree, glaciers grow the most where the sun shines least. Seemingly glued to the north sides of mountains, more than seven hundred glaciers are scattered about the North Cascades, more than anywhere else in the lower forty-eight states. But they are disappearing. The number of glaciers in North Cascades National Park is dropping (from 316 to 312 in recent years), even though it can snow any month of the year. Recent casualties include the Lewis and Lyall Glaciers at Rainy Pass.

Glaciers are not ordinary ice. Glaciers sustain life, support irrigation, and create and destroy ecosystems. This is ice that leisurely carves mountains into parapets and dust as we go about our daily lives. Glaciers are ice with attitude. But if glaciers disappeared, mountains would still retain their beauty, mystery, and majesty. Would glaciers' disappearance really matter? Who would notice besides the few who find entertainment climbing with spikes on their feet in unforgiving worlds? Very few would notice. (Out of sight, out of mind.) But most of us living on either side of the Cascade Mountains would be affected. And ice worms would notice.

Glaciers are an essential part of mountain ecosystems. There are even ecosystems *on* glaciers. Cold-tolerant extremophiles include watermelon algae (they have a pinkish, sun-protecting pigment), bacteria, spiders, microscopic fungi, hundreds of species of insects, and ice worms (which feed on the bacteria and algae). In one North Cascades glacier, the population of these one-inch distant relatives of earthworms was estimated to be greater than that of people on the planet. These snow and ice communities are a small link in the atmospheric carbon process and nitrogen cycling, and a food source for birds, such as rosy finches and ptarmigan, linking this hypothermic environment with other ecosystems.

Plants and animals downstream, too, depend on glaciers. When summers are hottest, survivability of

Glacial meltwater keeps streams flowing.

salmon, trout, and amphibians such as the Pacific giant salamander and the tailed frog rely on glacial runoff that keeps streams flowing. Variations of just one degree in average temperature can cause glaciers to grow or shrink, and a reduction in ice means less meltwater to sustain stream flows. Observations made below the Lewis Glacier indicate a 70 percent reduction in August streamflow after the loss of the glacier.

Farther downstream, urban ecosystems (known to be inhabited by mechanics, artists, nurses, teachers, and other essential organisms), too, depend on glacial runoff. About 90 percent of Seattle's electricity comes from hydroelectric generators kept spinning by the kinetic energy of moving water (not all of it glacial). Even our apples, peaches, and pears are glacier dependent. Grown on the dry east side of the Cascades, these fruits absorb molecules of life-sustaining irrigation water that flows down from alpine cirques. Decreased glacial mass leads to increased costs in modern living.

Sensitive to small changes in climate, glaciers reveal interconnections among energy, matter, and life. Glaciers intrigue us, buoy the human spirit, cradle colonies of cold-loving organisms, and feed life down-valley with that very rare substance in the solar system, water.

Remote, out of sight and out of mind, these lethargic rivers of ice are a part of the wide, interwoven network of life on earth.

Watermelon algae

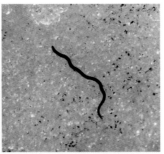
Ice worm (*Mesenchytraeus solifugus*), about one inch

Geologic Maps

I WAS TRYING TO FIND the transition between pink and green. After driving down from Washington Pass, which parallels eight miles of Golden Horn granite, I knew it would be around mile 170. There was no clear line, but I found the transition from crystalline granite to layers of sedimentary rock near mile 171. It had to be there, because I saw it on a Washington State geologic map.

Road maps show transportation routes. Weather maps show temperature, pressure, and precipitation. Geologic maps show rock, and any good geologic story begins with rock. What happened in this place? Were these rocks formed under the sea? Have they traveled far? How did they get here? I found fifty-million-year-old gas-rich magma bordering hundred-million-year-old coastal North America. It was the line between pink and green.

A geologic map shows locations, ages, and types of rocks, as well as structures such as faults, in kaleidoscopic form. Shaded yellow, sediments from glaciers and rivers fill the valley from Rockport to Puget Sound, one gateway to our North Cascades journey. Driving east, I find purple-and-gray lumpy tongues of Eldorado Orthogneiss and Skagit Gneiss bordering a pink wedge of Black Peak and Golden Horn granites before giving way to the green east-side sedimentary rocks.

Colors vary on geologic maps; typically reds indicate once-molten rock, and blues or greens point to sedimentary rock, but this is a convention more than a rule. Colors represent units of rock. Similar rocks may be colored in different shades, transitioning perhaps from burgundy to dark purple to lilac. What emerges is a picture of relationships telling which rocks formed where and when. Geologic maps may indicate mineral or ore deposits (from gold to gravel), tectonic histories, or ancient seas. They show structural elements such as faults, folds, and the dip angle of rock layers, clues to groundwater-flow directions and the likelihood of landslides or earthquakes. From the surface geology, we can make cross-sectional slices of the earth that provide another perspective.

But however beautiful a geologic map may be, its accuracy is only as good as observations made on the ground. Field geologists walk the ridges and valleys, carefully recording rock types, directions, angles, and locations. They collect and analyze specimens. They use their hand lenses in the field and polarizing microscopes in the lab. In the end, the map is a tool for interpreting the landscape. With this tool, scientists and hobbyists can go back in the field to look for minerals, decipher geologic hazards, solve water resource issues, or satisfy the human desire to understand a place.

The father of geologic mapping was the English surveyor William Smith. He spent years walking the countryside making detailed observations and collecting fossils. His epiphany that rocks and fossils were layered in predictable patterns led to his 1815 creation of a color-coded map of rock units stretching across Great Britain. North Cascades geologic mapping began with the U.S. Geological Survey geologist George Otis Smith in 1901. In the 1960s, University of Washington professor Peter Misch led cadres of future geologists through the rugged landscape. U.S. Geological Survey geologic mapping in the area culminated with the publication of the regional *Geologic Map of the North Cascade Range*, by Ralph Haugerud and Rowland Tabor in 2009.

THE THREE MAJOR GEOLOGIC DOMAINS CROSSED BY THE NORTH CASCADES HIGHWAY, WITH SELECTED PEAKS

WESTERN
Very old and very young ocean-floor sediments and volcanic rocks

METAMORPHIC CORE
A complex mix of deformed ocean sediments, volcanics, and magma intrusions

METHOW
Ocean sediments from a shallow sea

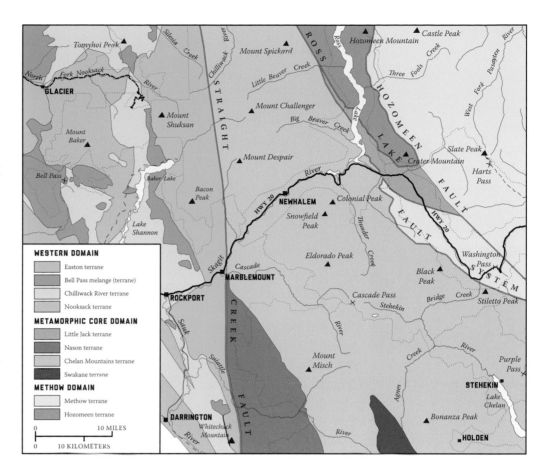

GEOLOGIC MAP OF WASHINGTON STATE

TAN
sediments (mostly nonglacial)

YELLOW
sediments (mostly glacial)

GREENS
sedimentary rocks

ORANGES
volcanic rocks

REDS
intrusive igneous rocks

PURPLES
metamorphic rocks

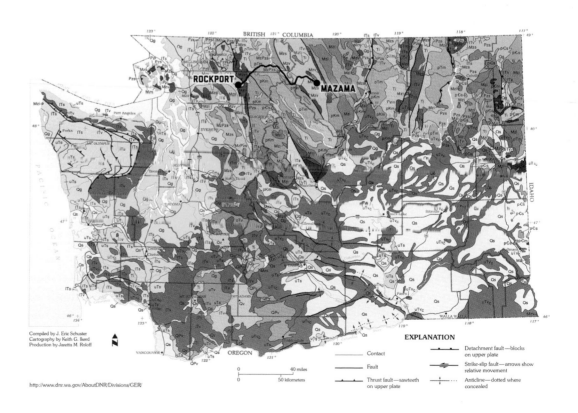

Geologic Time

Great ranges of mountains have risen up and worn away;
the lives of all our ancestors have flashed by.—Robert Jaffe

HOW TO COMPREHEND the incomprehensible? If you are bewildered by the thought of a whole mountain washing to the sea, grain by tiny sand grain, you begin to comprehend geologic time. It's difficult to accept this vast time span. To young eyes, a sixty-year-old grandparent is very old. The Revolutionary War is well beyond our lifetime. The Great Pyramids? Ancient history. Beyond that it all seems the same. Is there much difference between the times of extinct saber-toothed tigers (ten thousand years ago), the first humans (three-plus million years), the last dinosaur extinctions (sixty-five million years), or the formation of a North Cascades rock (ninety million years)? All these happened a very long time ago.

How to comprehend this "deep time"? Start with one piece. Start with a stream—look close, on the bottom. Grains of sand bounce and scatter over each other, slipping away. How long until they reach the sea? Days. Or years. Possibly many years. *Eventually* thousands of tiny sand grains reach a bay or shore, joining billions of others washing back and forth on the beach. Eventually. How many years until trillions of grains pile up thousands of feet thick in layers under so much pressure that water squeezes out and sand compresses and cements into sandstone? How long until a tectonic plate, moving a few centimeters per year, shuttles that sandstone toward the upper mantle of the earth, where it metamorphoses as it is pulled deeper? All of humanity is a blink. How long until compression from colliding plates slowly raises the metamorphic rock in a coastal mountain range eight thousand feet higher, and hundreds of miles from, where the sand grains once skittered on a creek bottom? It's close to incomprehensible. And the story doesn't end. On high peaks everywhere, ice and water and gravity break and carry hillsides and stone fragments, tumbled into sand grains, to creek bottoms, where they slip downstream toward the sea.

NO BEGINNING OR END?

Scottish farmer James Hutton picked up rocks in the late 1700s and wondered about their origins. He didn't know about plate tectonics, but he saw sand

Former sea-bottom sediments on top of Sourdough Mountain. How long until grains of this rock meet the sea again?

grains and sandstone and intrusions of rock. His mind saw cycles from mountain to sea, the mountains rising and wearing down again. He concluded that subterranean heat (probably from coal) caused mountains to rise and then they'd decay. Hutton understood the earth must be ever-changing and far older than what accepted scriptural doctrine told him. He declared, "We find no vestige of a beginning, no prospect of an end," a controversial and paradigm-shifting idea.

Hutton wasn't antireligion however. To the contrary, he, like Darwin, was deeply religious. That there was no vestige of a beginning and no prospect of an end reflected Hutton's view that the world was divinely timeless, self-perpetuating, and cyclical. He didn't know that one day we'd discover radioactivity and have a way to date rocks and the earth.

Hutton's observations and ideas led to the foundational idea of modern geology, that "the present is the key to the past." By observing current-day erosion, deposition, volcanoes, glaciers, and earthquakes, we can interpret the past. With modern technology, scientists measure subtle earth movements, magnetic orientations of rock, and proportions of radiometric isotopes, determining precise ages and locations of past events. The oldest rocks on earth? They are 4.3 billion years old.

RELATIVE OR ABSOLUTE?

In life and geology, events are sequential. You were born after your parents were born. One event is younger, the other older. This is *relative* time. But you have a specific birthdate, an *absolute* time. Both are important references in geology.

Superposition means the younger layers are on top (mile 146.5)

SUPERPOSITION

The Grand Canyon is an open cut in a giant recycling bin. The youngest layers were deposited on top of progressively older layers. The same is true elsewhere: flood deposits, landslides, lake-bottom mud, wind-blown sand, glacier-dumped gravel. Layers that are now on top had to be deposited after the bottom layers were already there. In 1669 Nicolas Steno described this relative age relationship as the law of superposition. This law is easily observed where sediments lie in simple layers, as in the Grand Canyon or the bluffs of Puget Sound. Stand at the base of the Tonto Platform or a Puget Sound cliff, reach out, and touch the oldest layers.

The North Cascades are a different story. Mountain building processes deform, melt, or overturn rock layers, complicating the picture. Rock outcrops and peaks reveal parts of the story, but the tale is so complex, stretching as it does over hundreds of millions of years, that geologists still disagree about pieces. Reading geologic clues in these mountains is like reading somewhat random pages from a novel: a few faded pages from chapter 3, pieces of chapters 5 through 8, and smudged pages from chapters 10 through 12; and then most of the end chapter rounds out the story. Moreover, some of the chapters have been rearranged. Even so, using principles of geology, chemistry, and physics, scientists can piece together much of the story. Fortunately, many of the page numbers are preserved as radioactive isotopes.

But continuity is disrupted throughout the North Cascades. Movement along fault zones displaces rocks by sometimes dozens or hundreds of miles. Rocks on either side of the Straight Creek and Ross Lake Faults have moved so far north or south that now-neighboring rock units have completely different ages and histories. On Jack and Crater Mountains, thrust faults pushed older seafloor on top of younger rocks. Molten intrusions forced their way into rocks all over the Cascades, displacing, baking, and recrystallizing their new neighbors and sometimes resetting their radiometric "birthdates."

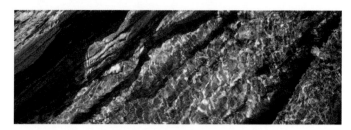

Which layer is on "top" in these tilted bands of streambed gneiss?

Because the North Cascades are not simple layers of sedimentary rock, it is mostly impossible to tell which rocks are older or younger simply by looking at them.

ABSOLUTE TIME

Atoms exist in different forms called isotopes. For example, ^{12}C and ^{14}C are forms of carbon with atomic masses of 12 and 14. Some isotopes are not naturally stable, so the atoms decay into stable forms of other elements. We use the rate this occurs to calculate the age of rocks.

Potassium (^{40}K) decays into Argon gas (^{40}Ar). When this happens in molten rock, the newly formed argon gas escapes the magma. But when the magma cools to hardened rock such as granite, a "clock" is set. In potassium-bearing crystals such as hornblende and mica, argon begins to accumulate because the atoms are trapped in the mineral's crystal structure. This $^{40}K/^{40}Ar$ decay happens at a very steady and specific rate, so geochemists measure the ratio of K/Ar to determine the age of the crystal and therefore the rock.

Another technique uses zircon (Zr). Zircon crystals often contain uranium (U) atoms in their crystal structures. Some isotopes of uranium (^{235}U and ^{238}U) decay to lead (^{207}Pb or ^{206}Pb), which remains in the zircon crystal. Like K/Ar, the U/Pb ratio is measured to determine the age of the rock.

Geologic Timetable

SELECTED ROCKS AND GEOLOGIC EVENTS SEEN ALONG THE NORTH CASCADES HIGHWAY

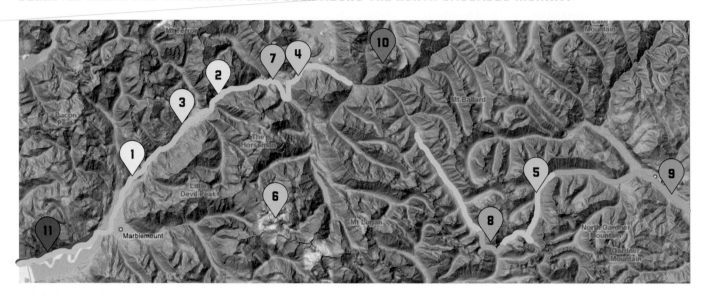

ERA	MAP LOCATION	APPROXIMATE AGE (Years) m = million	EVENT	MILE MARKER LOCATION
CENOZOIC	1	12–15,000	Glaciers, then rivers, fill valley bottoms with sediment.	approx. 60–110
	2	12–15,000	Skagit Gorge is cut by the release of glacier-dammed water.	121–124
	3	2.2–34 m	Chilliwack Batholith: granitic magma invades the Cascades; this large mass was once internal plumbing of volcanoes that eroded away long ago.	114–119
	4	45 m	Skagit Gneiss Complex: crustal extension and uplift cause metamorphism.	121–138
	5	45–48 m	Golden Horn Batholith: granite with high concentrations of rare minerals intrudes near the surface of the earth's crust.	144–156, 160-171
MESOZOIC	6	75–90 m	Eldorado Orthogneiss: magma invades older rock; later it is squeezed, stretched, and uplifted, producing Eldorado Peak.	100 (view point)
	7	90 m	Skagit Gneiss Complex: plate collisions cause intense metamorphism with heating, squeezing, and stretching.	121–138
	8	90 m	Black Peak Batholith: magma intrudes older, metamorphosed ocean sediments and volcanic rocks.	154–160
	9	100–110 m	Winthrop Sandstone: sediments and vegetation are deposited in broad meandering river plains.	171–190
PALEOZOIC	10	250–320 m	Hozomeen Terrane: deep ocean sediments and lava form; much later, they are thrust onto the continent, becoming the upper half of Jack and Crater Mountains.	approx. 141 (does not directly border highway)
	11	170–375 m	Chilliwack River Terrane: seafloor sediments and lava form along a volcanic island chain; much later, they are folded and faulted while colliding with the continent. Sauk Mountain is composed of these rocks.	90–103

THE NORTH CASCADES are a geologic jigsaw puzzle. To help make chronological sense of the puzzle, these two pages display representative geologic events seen along the highway. Time is not consistent across the mountains, owing to the extensive faulting, magma intrusions, and terranes from different places that are all tectonically sutured together. The complete geologic picture is substantially more complicated than presented here (see Appendix B).

Mile 108: Diobsud Creek. The Skagit Valley is filled with glacial sediments; then the ice retreats.

Mile 121: Skagit Gorge is gouged out by broken ice-dam floodwaters.

Mile 116: The forest takes over Chilliwack Batholith granite, former volcanic plumbing.

Mile 131.7: Diablo Overlook. The Skagit Gneiss Complex is created when rock is heated, stretched, and partially melted.

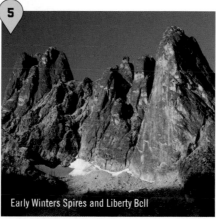

Early Winters Spires and Liberty Bell

Mile 162: Washington Pass Overlook. Golden Horn Granite is magma with unusual chemistry.

Eldorado Peak

Mile 100: Eldorado Orthogneiss is the result of a magma invasion, squeezing, and stretching.

Mile 183: Winthrop Sandstone is a result of river-plain plant fossils.

Mile 141: Crater Mountain. Hozomeen Terrane features seafloor pillow lava.

Mile 95: Sauk Mountain is built from former volcanic island sediments.

Plate Tectonics

One reads a palimpsest, not an essay. Time has erased, snipped, covered, twisted, and rearranged the text. —James Martin

THE NORTH CASCADES TEXT is rearranged. Southern rocks migrated north, western rocks moved east. Seafloor lava rests on mountaintops, and former sediments entwine with once-molten rock. This geologic mash-up has something in common with a covered pot of spaghetti sauce: heat finds ways to escape—and sometimes moves things in the process.

Earth is cooling off. It's so hot inside that the outer core is molten. Through conduction and convection, heat breaks out, in places bursting through vents and creating volcanoes and hot springs. But some of earth's internal heat is transformed into the kinetic energy of moving tectonic plates. Slow, unceasing movements in the earth's mantle force the more brittle upper mantle and crust (the lithosphere) to fracture and move, but at centimeters per year. Cracked into jigsawlike pieces, these lithospheric plates separate, slide, and run into each other, forming many of earth's features: The Rocky Mountains. Mount Saint Helens. All of Japan.

Where plates drive into each other, the earth wrinkles and fractures, spawning the world's largest mountain ranges. India runs into Asia. Africa into Europe. The Pacific Ocean floor slips under South America. The Himalayas form. The Alps. The Andes. Sometimes microplates of ocean floor barge into continental margins. Sediments divide like bread dough, some squeezed under the continent to bake and morph, and some, such as the Olympic Mountains, scraped and plastered on top of the continental pile. Some lithospheric plates slide and grind *past* each other: the San Andreas Fault. And plates separate, erupting the largest feature on earth, the forty-thousand-mile series of volcanic midocean ridges that wrap the earth like seams on a baseball. Iceland is the largest visible feature of this colossal hidden mountain range.

Foreground: near Teebone Ridge. Background: the Picket Range. Both ranges formed from tectonic forces pushing up earth's crust. Photo taken from Hidden Lake Peak.

The Cascade Mountains formed as Pacific plates migrated east, colliding with the massive North American Plate. The Kula ("all gone" in native Tlingit) Plate was entirely consumed. The once-mighty Farallon Plate shrank to become today's diminutive Juan de Fuca Plate, which will ultimately vanish like the Kula, recycling back into the earth. Plate collisions sutured volcanic island chains and ocean crust onto North America. Along with north-south fault movements and rising magma from below, the earth's giant heat-releasing mechanism rearranged the Cascades like cubist Picasso paintings. Convergence continues. Today, the Juan de Fuca Plate subducts beneath Washington one to two inches per year and will continue to drive mountains upward, trigger earthquakes, and stoke volcanic eruptions.

Erosion

The rhythm of the rocks beats very slowly, that is all.—Colin Fletcher

WHILE FORCES IN THE EARTH spit out lava and push mountains skyward, gravity teams with wind, water, and ice to bring it all back down. With enough time, whole mountain ranges wash to the sea. A Himalaya-size range existed four hundred million years ago, running from the area where the Appalachians now stand, through Nova Scotia's Cape Breton Highlands, the Scottish Highlands, and the Jotunheimen Mountains of Norway. As tectonic forces spent the last two hundred million years widening the Atlantic Ocean, the ancient mountains have been slowly eroding, filling in the edges of the ocean that split the range apart.

Gravity is the unrelenting twenty-four-hour-a-day erosion captain. Trying to smooth the earth to a billiard ball, gravity pulls highs to lows. Planets and stars are round because gravity always wins. Even ski posters slyly say, "Obey the law," because everything must go down.

It's not just sand grains that make their way to the sea: gravity brought this massive boulder down in a landslide at mile 122.

THE POWER OF WATER

Gravity doesn't work alone, though. Water cocaptains the erosion team—it would be unbeatable in the game of rock-paper-scissors. Each item can beat another, but add water as a choice and the game is over. Water fractures rock. Dissolves paper. Rusts scissors. Water beats them all. As water freezes, it expands, splitting rock into smaller pieces. Gravity topples the newly unbalanced rocks. Water carries the fragments toward the sea (pulled by gravity), its streams joining together like airport conveyors whisking luggage away. The Skagit River carries an average of more than seven thousand tons of sediment into Puget Sound every day.

Water expands when it freezes. This rock was split by wedging ice. Photo taken at the top of Mount Pilchuck, Washington

Water does more than transport though. Carbon dioxide, sulfur, and nitrogen from the air dissolve in water to form weak acids that chemically alter certain minerals. Feldspar, the most common mineral in the earth's crust and a key compound of many igneous and metamorphic rocks, eventually weathers to soft clay. Hard quartz pieces are left behind, which break into sand grains. Gravity, wind, and water then do their job, shipping the particles from mountain to sea. Erosion is persistent if nothing else.

The North Cascades are home to an exceptionally erosive form of water: glaciers. Mount Baker alone submits a massive amount of sediment to Puget Sound. Glaciers tenaciously scrape rock into powder and even provide meltwater to flush out and haul the sediments away.

It's all crumbling. Right before our eyes. But don't expect a new scene soon. It takes ultimate patience, reducing a mountain grain by grain.

Rocks

The stones are the bones of the beast that rolls in the rapture of orogeny. —Robert Michael Pyle

ALL ROCKS have been something else. Especially in the North Cascades. Sandstone, basalt, marble, granite, amphibolite: they are all recycled or altered. Change is the constant. After rocks form on a volcanic slope, ocean bottom, or subterranean magma chamber, conditions change. The rocks are exposed to air, rain, running water, or ice. Gravity. Increased pressure or decreased pressure. Or changes in temperature that can recrystallize mineral grains and alter their colors. Certain rocks form in certain conditions. When conditions change, so does the rock. The slow arrangement of atoms into perfectly ordered geometric crystals is a change toward the lowest energy condition.

Gray limestone can metamorphose into elaborately textured marble. Gray shale can transform into shimmery schist, its iridescence like fish scales frozen in sunlight. But most rocks—granite and gneiss, tonalite and talc, greenstone and blueschist—will end up as sand. The sand will become sandstone, and another cycle will begin; burying, heating, foliating, lineating. Metamorphism, tectonism, plutonism. Rocks melt, cool, crystallize, ride in tectonic plates, and rise in another place far from here in geography and time.

If magma contains 72 percent silicon dioxide, 14 percent aluminum oxide, and smaller percentages of potassium, sodium, calcium, iron, and magnesium oxides, and the magma cools slowly underground, it will form granite. If the magma has minute percentages of lithium or manganese, purple minerals may form: lepidolite, spodumene, phlogopite, purpurite. With magnesium or iron present, clear quartz becomes purple amethyst or yellow citrine. Potassium yields salmon-colored orthoclase feldspar. But if this same magma, with all its potential for colorful crystals, extrudes aboveground, it cools quickly into gray lava. As in real estate, location is everything.

In middle and high school earth-science classes, students study the Rock Cycle. Diagrams with arrows connect igneous, sedimentary, and metamorphic rocks in a circle, with other stages and processes in between: sediments, magma, erosion, melting. It's hard for a diagram to convey this complex reality, so teachers in the Puget Sound region may take their students for a firsthand look—to the beach! There, igneous and metamorphic rocks are in abundance because Pleistocene glaciers have indiscriminately left their rock collections carried from afar. Though the rocks are all rounded and smooth, each color and grain structure tells a story about chemistry, temperatures, and conditions far different from those of their shared, and *temporary,* shoreline home.

The North Cascades are a tangible medley of the rock cycle, seemingly frozen in a human time line. These rocks seem permanent as we walk, bike, or drive past, or slowly climb their walls. What takes thousands or millions of years is mostly invisible to the human eye.

shale   schist

All rocks are changing. These roadside rocks are oxidizing, turning black. Chemically weakened, the rocks are more prone to breaking apart in response to freezing water. Gravity pulls a weak zone down. With more surface area exposed, more of the rock oxidizes, becoming weaker. Slowly but surely changing.

The granite of Early Winters Spires formed deep underground, hot and molten. Temperatures cooled, minerals crystallized, and tectonic pressures forced the rock skyward. Now the cliffs are slowly transformed by the sun, rain, wind, ice, and by gravity. Once-molten magma became mountain peaks that will turn to sand, to be carried to the faraway sea.

Igneous Rocks

Lava cools and hardens into igneous rock at Krafla, Iceland.

GRANITE, foundation of skyscrapers, and basalt, foundation of eastern Washington, come from molten origins.

Igneous rocks form either from magma (below-ground) or lava (above). Tourists delight when rivers of basalt flow across land in Hawaii, Iceland, and other volcanic regions—basalt is an *ex*trusive rock, easily seen when molten, snaking across the landscape burying roads, destroying homes, or burning forests. But you won't see molten granite flowing anywhere. Granite cools and crystallizes underground. It is an *in*trusive rock. To see it, you'd have to travel two to twenty or more miles into the earth's crust.

Igneous rocks differ not only in *where* they form but also *what* they form from—their composition. More silicon, with less iron and magnesium, makes lighter-colored rocks such as granite or a high-silica lava called rhyolite. The opposite equals darker rocks such as basalt. The Cascade mountain region tends to occupy the middle of this scale, so in addition to granite there is medium-silica diorite and tonalite. Instead of low-silica basalt, lava tends to be dacite or andesite (or sometimes even rhyolite). Mount Saint Helens has erupted basalt, andesite, and dacite lavas; its current lava dome is dacite.

Rhyolite (high-silica lava) at Slate Peak

GRANITE MASSES

When tectonic plates collide, immense pressures create folded and faulted mountain ranges. Deep in the crust, rocks melt. Molten granite (or diorite or tonalite) can form in miles-wide bodies called batholiths. Uplifted and eroded, these masses form the foundation of major passes in the North Cascades: Snoqualmie, Stevens, Rainy, Washington.

Magma chambers can take thousands of years to cool, so there's plenty of time for molecules to arrange themselves into perfect geometric shapes. Hexagonal quartz crystals. Rhombic feldspars. Dodecahedral garnets. But *ex*trusive rocks cool quickly into small, less visible grains. And some magma cools in two stages, producing a porphyry—crystals in a fine-grained matrix.

Lacking the crystalline beauty of their intrusive cousins, extrusive igneous rocks have their own interesting properties. Differences in magma composition, temperature, and gas pressure produce glassy obsidian, featherweight pumice, volcanic bombs, fine-grained ash, porphyritic tuff, ropy pahoehoe, and skin-piercing aa.

Blankets of pumice (extrusive) cover peaks of granite (intrusive) at High Pass, east of Glacier Peak.

Granite formed with pockets of quartz crystals near Granite Pass.

This pumice was excavated by a burrowing marmot.

Quartz and plagioclase feldspar

Golden needle of astrophyllite

Arfvedsonite and quartz

Microscopic crystals in Golden Horn granite gas pockets: each mineral forms a different geometric shape based on the orderly arrangement of its atoms.

GRANITE

The word *granite* may evoke images of colorful interlocking crystals in kitchen countertops, but most granite is fairly plain. Much of what people call granite is black-and-white granodiorite, including the famous El Capitan and Half Dome cliffs of Yosemite National Park. In the North Cascades, much of the "granite" is tonalite. The compositions of granite, granodiorite, and tonalite vary mostly in three elements: potassium, sodium, and calcium. (Fans of the periodic table, notice where these elements are located.) Like all granites they form up to thirty miles underground from slow-cooling masses of magma. Sometimes, gas pockets remain in the hardened rock, leaving cavities where crystals grow. Crystals vary in size, from microscopic (collected as prized "micromounts") to many feet long, including an eighteen-foot beryl crystal found in the state of Maine.

GOLDEN HORN GRANITE

The "Golden Horn" has minicavities with tiny crystals of quartz, feldspar, and rare minerals, from aenigmatite to zektzerite. The Golden Horn is a low-calcium alkali granite. The forty-eight-million-year-old gas pockets formed when water boiled out as the magma rose and cooled just two miles below the surface. Golden Horn granite makes up Early Winters Spires, Cutthroat Peak, the Needles, and other mountains of the Washington Pass area, including, of course, the peak simply named Golden Horn.

GRANODIORITE AND TONALITE

Most granite in the north Cascades is granodiorite or tonalite, granite with more calcium and less potassium. The difference in composition affects the minerals and look of the rock. The granite at Stevens and Snoqualmie Passes is mostly granodiorite. Many of the gneisses in the Skagit Gneiss Complex derive from tonalite.

Metamorphic Rocks

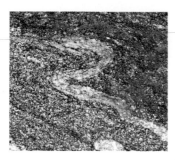

METAMORPHIC ROCKS are survivors. Pick one up and appreciate the turmoil it's been through, escaping total mineral meltdown near a magma chamber of two thousand to three thousand degrees Fahrenheit. Or bathing in super-heated geothermal liquids. Or adjusting to the vicelike squeeze of tectonic plates fifteen miles below the earth's surface. Metamorphic rocks are, by definition, changed rocks (*meta* means "form"; *morph* means "change"). They were originally sedimentary, igneous, or other metamorphic rocks. Some have transformed from one type of metamorphic rock to another, some are so greatly altered that their origins cannot be identified, and some are barely changed at all.

It's not always clear if you're looking at a metamorphic rock. A sure sign is that it looks foliated. Layers can be a clue, but look closely with a hand lens to see if the grains are cemented together (sedimentary rock) or are interlocking crystals (metamorphic). (Igneous rocks, too, can have visible interlocking grains but aren't layered).

One typical metamorphic rock sequence starts with shale. Mud accumulates in ponds, lakes, or the ocean abyss. If buried deep enough, mud compacts to shale. With depth come higher temperatures and pressures. Shale compresses to slate. (Denser than shale, slate dings with a higher pitch when tapped with a hammer.)

Next, phyllite. Clay minerals in slate crystallize to layers of silvery mica, which shimmers in the light. Phyllite morphs to schist. Mica grains enlarge. With enough time, schist evolves to gneiss. With the right chemistry, temperatures, or pressures, atoms rearrange to new minerals: black biotite mica, cross-shaped staurolite, or diamond-sided garnet crystals. If the rock melts, its mineral structure is erased entirely. It becomes liquid. Magma then cools to the igneous stage of the rock cycle.

Any rock can be metamorphosed. Bury it deep and wait. Ancient algae mats became limestone, which became marble, now visible near mile 127. Sandstone becomes quartzite. Basaltic lava becomes greenstone or amphibolite. The best-known North Cascades metamorphic assemblage is the Skagit Gneiss Complex, roughly surrounding the road from mile 121 (Skagit Canyon) to mile 135 (Ross Lake). These gneisses formed from granitic rocks and ocean sediments. There's banded biotite gneiss. There are paragneisses, orthogneisses, metagraywacke gneiss, and even hornblende-cummingtonite-garnet gneiss. It's gneiss heaven for metamorphic-rock hounds. And there's migmatite, a mix of gneiss and melted or partially melted granitic rocks—a hybrid of the igneous and metamorphic worlds.

An easy location to see metamorphism on the North Cascades Highway is across from the Diablo Lake overlook, mile 131.7. Based on structure and mineral composition, this orthogneiss is a migmatitic hornblende-cummingtonite-garnet gneiss!

Biotite (black mica) schist

Shuksan Greenschist: metamorphosed ocean basalt

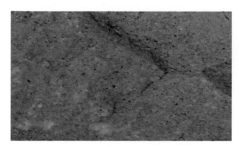

From sandstone . . .

. . . and from granite . . .

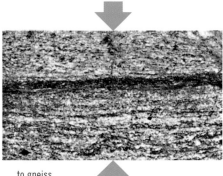

. . . to gneiss

GNEISS

Of quartz there are many rock puns. In metamorphics, it's about the gneiss schist. But there *is* nice schist, and there is nice gneiss. Choice specimens hold unusual minerals or aesthetic curves and foliation. Gneiss has a layered texture. With high pressures and temperatures (more than eleven hundred degrees Fahrenheit, six hundred degrees Celsius), elements segregate and align in bands of light minerals (mostly quartz and feldspar) and dark minerals (mostly biotite and hornblende). Give granite or sedimentary rock enough pressure, time, and temperature, and you have gneiss.

The hills around Gorge, Diablo, and Ross Lakes are Grand Central Station for the Skagit Gneiss Complex. These rocks have been through more than two hundred million years of sediment deposition, plate collisions, igneous intrusions, faulting, folding, more intrusions, more heating, more bending, more breaking, and more rearranging.

Orthogneiss. This molten rock cooled deep underground, forming granite, granodiorite, or tonalite. Later it was squeezed and layered into gneiss.

Schist

Banded gneiss. Layers of orthogneiss alternate with layers of schist. In the Skagit Gneiss Complex, the schist tends to be older, the orthogneiss came from younger intrusive rocks.

Migmatite on Sourdough Peak; Ross Lake is in the background.

Distorted feldspar crystals in gneiss, near Gorge Lake

Sedimentary Rocks

SEDIMENTARY ROCKS bookend the North Cascades' core. The Western Domain and the east-side Methow Domain rocks are mostly made from sand, pebbles, and mud mixed with volcanic rock. Between the sides, the Metamorphic Core Domain also has many rocks of sedimentary origin, but as the name implies, most of those have been altered to something else.

Conglomerate, sandstone, and shale are sedimentary rocks—collections of pebbles, sand, or clay deposited where the kinetic energy of moving water, wind, or ice slows to a minimum. Gravity is the great unavoidable force bringing everything down, down, until there is nowhere lower to go.

Swept to the sea, tumbled and rounded, pebbles, sand, and mud, combined in a slurry, start over as *conglomerate*. Angular, less traveled fragments cemented together make *breccia*. Where sand is deposited in large quantities, natural calcium or silica cement binds the grains into *sandstone*, which is sometimes so solid it is used as foundation stone for skyscrapers. But the finest particles travel until they reach calm water—a lake or sea bottom. Eventually, the silt and clay may compact into smooth and fissile *shale*.

Conglomerate, Midnight Peak Formation, Robinson Creek

Breccia, from Washington Pass

On the east side, layers of sandstone, conglomerate, shale, and volcanic rock near Mazama

On the west, a limestone quarry near Concrete

Sandstone, Winthrop Formation, Mazama

Shale, Chuckanut Formation, Bellingham

Then there's *limestone*. Ocean water contains dissolved calcium carbonate, from which clams, mussels, and other organisms build their exoskeletons. Limestone forms from precipitated calcium carbonate at the bottom of the sea, sometimes becoming so uniform and solid that it, too, can be used for foundation stone (but it is prone to chemical weathering). Limestone also forms from skeletal fragments of millions of diminutive marine organisms that collect on the seafloor.

The town of Concrete, in Skagit County, owes its existence to 330-million-year-old ocean-bottom lime. Thrust to the surface as a wedge of ocean floor, marine sediment became the raw material for Concrete's cement-making industry.

Limestone, Chilliwack River Formation Concrete

FOSSILS

There's something about touching a fish in stone. Or splitting a rock and viewing two halves of a fern that hasn't seen light in forty million years. I remember finding my first trilobite. My first *Sigillaria*. Both in beach cliffs. I tried to imagine an ocean full of potato-bug-like creatures and a forest of giant treelike plants with lizard-skin bark. What else lived there? Why did they disappear? Who are their descendants today?

Fossils are found on both sides of the Cascades but not in the Metamorphic Core, where heat and pressure destroyed any possible remains. On the east, Winthrop is home to the Winthrop Sandstone and other sedimentary

ronments that preserved them. Marine and wetland muds shield organisms from decay and scavengers, so to be found a hundred million years after you die, it helps to live in a quiet aquatic environment. How many mountain, desert, or prairie species will never be known?

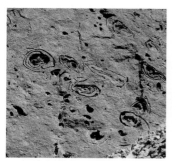

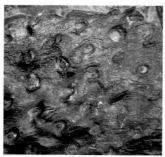

Snails and clams in sandstone, approximately 100 million years old, Virginia Ridge Formation, Slate Peak

Sigillaria bark at Moose River, Nova Scotia

Ferns in sandstone, 100 to 110 million years old, Winthrop Formation, Mazama

Palm leaf in shale, 45 to 48 million years old, Chuckanut Formation, Bellingham

formations with marine organisms and plants, which were buried in shallow seas and on a broad coastal plain with meandering rivers.

On the west side, near Bellingham, are subtropical plant fossils of the Chuckanut Formation. Fifty million years after the Winthrop fossils formed, Chuckanut river sediments preserved palm leaves, sequoias, mollusks, and bird and mammal tracks. Even a soft-shelled turtle. Near Concrete are much older, 330-million-year-old limestone deposits with crinoids, plantlike sea urchin relatives known as sea lilies. Near Mount Baker, a silt-stone formation full of marine fossils is appropriately named Chowder Ridge.

All the fossil species ever found are only a fraction of what actually lived. Most species (including dinosaurs) will never be known, because they didn't live in envi-

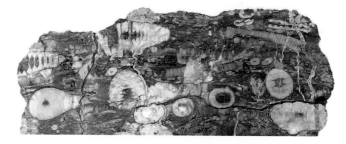

Crinoid pieces, more than three hundred million years old, in Chilliwack limestone, from the geology department collection at Western Washington University

45–48-million-year-old leaves in shale, Chuckanut Formation, Bellingham

Geologic Detectives

If an unfortunate lover of nature was seen hammering in a stone quarry, he was generally supposed to be slightly demented.—Peter Bellinger Brodie, 1858

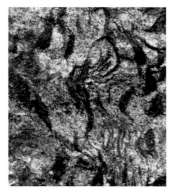

Polywog agmatite

DESCRIBING North Cascades rocks, the geologist Ralph Haugerud writes in *Geology of the Metamorphic Core,* "Synorogenic plutons are mostly tonalitic." This reminds me of Lewis Carroll's poem "Jabberwocky." The difference is that *tonalitic synorogenic plutons* actually means something. (It should be noted that when Dr. Haugerud speaks to the public, he describes geologic concepts in simple terms, with stories that bring geology alive.) North Cascades rocks include metaplutonic migmatite and granoblastic biotite, and at mile 139.9, you can see the polywog agmatite. *Polywog* makes sense when you see the rock looking like a mass of tadpoles swimming upstream. And agmatite? Well, *this* agmatite is a leucocratic trondhjemite, which contains oligoclase-based plagioclase.

GEOFORENSICS

Geologists are like forensic scientists, deciphering clues in the smallest details, so language needs to be precise. There is a difference between *orthoclase* and *plagioclase. Quartz* and *quartzite.* Names can divulge chemical details or clues to formation temperatures and pressures that help solve the riddle of how the landscape formed. Near Washington Pass I have enjoyed finding miarolitic cavities (small, irregular, crystal-lined pockets) and rapakivi texture (pink feldspar surrounded by white feldspar), which tell me that Golden Horn granite formed closer to the surface than, and has a composition different from, other Cascades granites.

Some geologists do actual forensic work. Murders have been solved with clues such as distinct rock fragments or microfossils in tire treads or shoe soles. The infamous 1960 kidnap and murder of the grandson of a beer company executive was solved when grains of pink Pikes Peak Colorado granite were found in the tires of the suspect's burnt-out car back in New Jersey.

GET OUTSIDE

Essential field tools

At heart, geology is a field science. The first geologists in this region were renowned for their wilderness skills. A Maine native and future director of the U.S. Geological Survey, George Otis Smith, investigated here in 1901. Author and climber Fred Becky describes Smith in *Range of Glaciers* as

a self-reliant explorer who was ready to venture for weeks into the untracked wilderness equipped with only the bare essentials: tools for measuring and mapmaking, fieldbooks, a geologist's hammer, and a good supply of bacon, beans, and coffee. His pack was always heavier upon his return, the food having been replaced by carefully labeled rock samples.

The father of North Cascades geology was the University of Washington professor Peter Misch. He and his graduate students climbed and hiked throughout the region while completing the first comprehensive geologic map of the area in the 1950s. His keen observations led to our understanding of the North Cascades as three distinct geologic regions. Some of his mentees are now the experts.

ICE AND SULFUR

Geologists' tools range from the classic essentials—hammer, chisel, and hand lens—to spectrophotometers, scanning electron microscopes, and ion chromatographs. Sometimes they even include steam drills and tank antennas. An ongoing Park Service study of North Cascades glaciers requires three data-sampling visits to remote glaciers per year. With a steam-jet hose, geologists slowly bore holes in the ice to place pipes they will use as measuring stakes. Snow depth on top of the ice is measured with World War II tank antennas. Days are split between a world of ice and rock and a world of fluorescent lights and computer screens. The data become part of a worldwide glacial ice data bank.

At even higher elevations, geologists get heat with their ice, sampling volcanic fumaroles in the crater of Mount Baker. Getting close and personal with steaming hot jets of sulfur-rich gas isn't for everybody, but it's the best way to understand what's happening inside the mountain. Gas samples are taken right from the source, then analyzed in a U.S. Geological Survey laboratory. These data, too, are shared with scientists and the public.

MORE THAN ROCKS

Geologists understand elements of physics, biology, astronomy, oceanography, meteorology, and chemistry and provide data about climate change (causes *and* effects). Such data can tell us how and where to grow plants. Where to find energy, mineral, and water resources. How to maintain healthy soils and clean water. How to prevent erosion. How mountain ranges form, when ice ages occurred, what caused extinctions. Geologists probe for warnings that signal volcanic eruptions and earthquakes. Analyze where colony-sustaining resources exist on the moon or Mars. Probe meteorites to understand the formation of the solar system. Geologists can be found in every corner of the planet solving earth riddles from the practical to the simply intriguing.

"JABBERWOCKY" REVISITED

And those tonalitic synorogenic plutons? They're cooled masses of magma with chemistry similar to that of granite, but they have less potassium and more calcium and were formed during the turmoil of tectonic folding and faulting. Simple.

A geology professor and graduate student study microfold structures in the Skagit Gneiss Complex.

National Park Service geologists collect snow-depth data at about 8,000 feet on the Silver Glacier.

Collecting volcanic gases on the floor of Mount Baker's Sherman Crater

Epilogue

NATIONAL PARK AND WILDERNESS LANDS IN THE NORTH CASCADES

Respect for preservation comes to a culture somewhat as it comes to a man. In his infancy, beauty or other intangible values do not register—he will smear an old newspaper or a rare painting with equal gusto. Some years later he at least respects his parents' evaluation of the painting and lets it survive. In maturity he will delight in a painting, know that it is worth far more than a yard of canvas, a pound of paint and wages. Perhaps it would be the last thing he would sacrifice.

. . . I think we have matured enough to seek commodities elsewhere and let the Northern Cascades masterpiece survive unspoiled for those who we should assume will surely mature enough to cherish it. We can save for them, in this wilderness, our greatest national park.
—Harold Bradley

AS PART OF A GROWING MOVEMENT, Bradley wrote this in 1958 to encourage creation of a North Cascades National Park. More than fifty years later, the movement continues because, back in 1968, enchanting areas like Blue Lake and the Washington Pass region were left out of North Cascades National Park. Change continues. As recently as 2012, Seattle City Light relinquished interest in a hydroelectric project at Thunder Creek, which led to 3,559 acres of that watershed becoming part of the Stephen M. Mather Wilderness. And there is still movement toward expanding the national park to include areas near Washington Pass, Cascade Pass, and Mount Baker.

SERENE PLACES

Wherever we seek to find constancy we discover change. Having looked at the old woodlands in Hutcheson Forest, at Isle Royale, and in the wilderness of the boundary waters, in the land of the moose and the wolf, and having uncovered the histories hidden within the trees

On the back side of Liberty Bell Mountain and Early Winters Spires, Blue Lake is a popular destination, easily accessible because of the North Cascades Highway. Visitors' respect for the area shows, because it remains pristine. This basin and the surrounding peaks are national forest land and so are not protected as part of the national park.

Concord Tower

North Early Winters Spire

Liberty Bell Mountain

Lexington Tower

South Early Winters Spire

and within the muds, we find that nature undisturbed is not constant in form, structure, or proportion, but changes at every scale of time and space. The old idea of a static landscape, like a single musical chord sounded forever, must be abandoned, for such a landscape never existed except in our imagination. Nature undisturbed by human influence seems more like a symphony whose harmonies arise from variation and change over many scales of time and space, changing with individual births and deaths, local disruptions and recoveries, larger scale responses to climate from one glacial age to another, and to the slower alterations of soils, and yet larger variations between glacial ages. —Daniel B. Botkin

I expect certain places to be there when I return. The high ledge, the still pond, the clear stream gurgling under dark cedars. These sites are hospices—comfortable, serene, familiar. I'm glad to find them each time I come back. They're never exactly the same, of course. That is part of the allure: seeing nature change. Trained as a geologist, though, I know that eventually they will disappear. Soil creep, landslides, volcanic eruptions, climate variations, forest fires, disease—in the end, nature will have its way. In millions of years mountain ranges will come and go, seas will rise and fade, and polar ice caps will grow and shrink. Even species will change. Extinctions will continue. New species will appear.

On this ledge of ancient migmatite, one of my refuges, one of my temples, I'm surrounded by national park and wilderness areas. Without these designations, the landscape might look very different. Highway 20 was built in part to bring state commerce and access to timber and mineral resources. In the late 1950s, promoters spoke of the wealth to be gained from vast mineral and overripe timber resources. I'm glad "overripe" timber still exists. When North Cascades National Park came to be, Mount Baker, the original crown jewel of the pro-

posed park, was excluded because of timber interests. Had it not been for strong-willed environmentalists, the ancient forests surrounding me would likely be patches of clear-cuts and monoculture tree farms. What is succulent, what is sweet, what is "ripe" would be gone. There would have been an open-pit copper mine in what is now Glacier Peak Wilderness. There would not be a North Cascades National Park. These magnificent forests with clean air, clean water, and diverse biological communities are natural, but they still exist because people fought for them.

The landscape *will* change. But there is value in preserving serene places. There is value in preserving water quality and minimizing erosion. There is value in helping species survive. We keep habitat for the grizzly and the rough-skinned newt because they are essential parts of a complex ecosystem—they are part of something much larger than ourselves. We protect clean water and air for salmon and eagle and us and our grandkids' grandkids. We humans are soulful, so we write songs and create art and keep unspoiled views unspoiled, and we delight in fairy slipper orchids, the deep woods of giant firs, and pure alpine lakes ringed with cliffs we'll never climb.

We can't preserve nature as a museum, but we can prevent egregious habitat destruction. We can avoid overharvesting. We can live with the paradox of managing wildlands to let nature take its course. We can act for future generations.

May we all take care of this place.

Climbers return to camp, Sahale Glacier

Wipperwil's in the Dawn.
Pretty soon he'll be gone.
But he's got one good song to sing.
But like my Daddy said.
It's in your heart not your head.
And you've got to sing and sing and sing.

—John Dillon and Steve Cash of
the Ozark Mountain Daredevils

The moon and Jupiter rise over serene Diablo Lake.

Conservation Organizations with Ties to the North Cascades

These organizations work to protect the long-term health of the greater North Cascades ecosystem.

NORTH CASCADES INSTITUTE Conserving and restoring Northwest environments through education. The North Cascades Institute seeks to inspire closer relationships with nature through direct experiences in the natural world. Located on the north shore of Diablo Lake. *www.ncascades.org/discover/north-cascades-institute*

NORTH CASCADES CONSERVATION COUNCIL Protecting and preserving the North Cascades' scenic, scientific, recreational, educational, and wilderness values. Home organization for the American Alps Legacy Project. *www.northcascades.org*

CONSERVATION NORTHWEST "Keeping the Northwest wild." Protecting and connecting old-growth forests and other wild areas from the Washington coast to the Rockies of British Columbia. *www.conservationnw.org*

NATIONAL PARKS CONSERVATION ASSOCIATION Protecting and enhancing America's national parks for present and future generations. *www.npca.org/about-us/regional-offices/northwest*

NATIONAL WILDLIFE FEDERATION Inspiring Americans to protect wildlife for our children's future. *www.nwf.org/Pacific-Region.aspx*

WASHINGTON NATIVE PLANT SOCIETY A forum for individuals who share a common interest in Washington's unique and diverse plant life. *www.wnps.org*

WASHINGTON WILD Protecting and restoring wildlands and waters in Washington State through advocacy, education, and civic engagement. *www.wawild.org*

A naturalist teaches students on a hike near Diablo Lake.

The following outdoors organizations, too, offer opportunities to participate in North Cascades conservation efforts:

THE MOUNTAINEERS *www.mountaineers.org*
WASHINGTON TRAILS ASSOCIATION *www.wta.org*
FEDERATION OF WESTERN OUTDOOR CLUBS (includes Washington and Skagit Alpine Clubs, Mount Baker Club), *www.federationofwesternoutdoorclubs.org*

Bibliography

Babcock, S., and B. Carson. *Hiking Washington's Geology*. Seattle: The Mountaineers, 2000.

Barksdale, J. *Geology of the Methow Valley, Okanogan County, Washington*. Olympia, WA: State Department of Natural Resources, 1975.

Becky, Fred. *Cascade Alpine Guide: Climbing and High Routes*. Vol. 2. *Stevens Pass to Rainy Pass*. 2nd ed. Seattle: The Mountaineers, 1996.

———. *Cascade Alpine Guide: Climbing and High Routes*. Vol. 3. *Rainy Pass to Fraser River*. 2nd ed. Seattle: The Mountaineers, 1995.

———. *Range of Glaciers: The Exploration and Survey of the Northern Cascade Range*. Portland: Oregon Historical Society Press, 2003.

Board of State Road Commissioners. "Final Report of the Board of State Road Commissioners of the State of Washington, 1896." *Washington Highways* (September 1972), www.wsdot.wa.gov/NR/rdonlyres/8D705E08-4DA8-425B-97ED-96ACE4E4B21A/0/WashingtonHwys_Sept1972.pdf.

Bourasaw, Noel, ed. *The Skagit River Journal,* www.skagitriver journal.com.

Bruggeman, Jason E. *Factors Affecting Pika Populations in the North Cascades National Park Service Complex, Final Report*. Farmington, MN: Beartooth Wildlife Research, 2011.

Darvill, F. T. *North Cascades Highway Guide*. Mount Vernon, WA: Reliance Printers, 1973.

Devin, D. *Mazama: The Past 125 Years*. Winthrop, WA: Shafer Historical Museum, 2008.

Edwards, John. "Lower Life in Higher Places." *Wild Cascades, Journal of the North Cascades Conservation Council* (Spring 2009): 7–8.

Figge, John. *Evolution of the Pacific Northwest: An Introduction to the Historical Geology of Washington State and Southern British Columbia*. Seattle: Northwest Geological Institute, 2009.

Fleischner, Thomas L., and Saul Weisberg. "Tidewater to Timberline, Forest to Steppe: Natural History of the Greater North Cascades Ecosystem." In *Cascadia Wild: Protecting an International Ecosystem*, edited by M. Freidman and P. Lindholt, pp. 4–21. Bellingham, WA: Greater Ecosystem Alliance, 1993.

Greenwald, Noah. *Petition to List the White-Tailed Ptarmigan (*Lagopus leucura*) as a Threatened Species under the Endangered Species Act*. Portland, OR: Center for Biological Diversity, 2010.

Haugerud, Ralph. "Geology of the Metamorphic Core of the North Cascades." In *Geologic Guidebook for Washington and Adjacent Areas*. Washington Division of Geology and Earth Resources Information Circular 86, edited by Nancy L. Joseph and others. Olympia: Washington State Department of Natural Resources, 1989.

Haugerud, Ralph, and J. Brian Mahoney. *Terrain Accretion along the Western Cordilleran Margin: Constraints on Timing and Displacement*. Field Trip Guidebook and Proceeding Abstracts, Penrose Conference, June 21–27, 1999, www.geology.cwu.edu/facstaff/huerta/g456-556/FT/Terrane%20Accretion%20along%20the%20Western%20Cordilleran%20Margin.pdf.

Haugerud, Ralph, J. Brian Mahoney, and Joe D. Dragovitch. *Geology of the Methow Block*. Seattle: Northwest Geological Society, 1996.

Haugerud, Ralph A., and Rowland W. Tabor. *Geologic Map of the North Cascade Range, Washington. Technical Pamphlet to Accompany Scientific Investigations Map 2940*. Denver: U.S. Department of the Interior, U.S. Geological Survey, 2009.

Hazen, R. M., D. Papineau, W. Bleeker, R. T. Downs, J. Ferry, T. McCoy, D. Sverjensky, and H. Yang. "Mineral Evolution." *American Mineralogist* 93 (2008): 1693–1720.

Jacobson, Daryl, Victor Pisoni, and Phil Woodhouse. *Discovering Washington's Historic Mines*. Volume 3. *The Northern Cascade Mountains*. Oso, WA: Oso Publishing, 2006.

Manning, Harvey. *The Wild Cascades: Forgotten Parkland*. San Francisco: Sierra Club, 1965.

———. *Wilderness Alps: Conservation and Conflict in Washington's North Cascades*. Bellingham, WA: Northwest Wild Books, 2007.

Martin, James. *North Cascades Crest*. Seattle: Sasquatch Books, 1999.

Mass, Cliff. *The Weather of the Pacific Northwest*. Seattle: University of Washington Press, 2008.

McGroder, Michael, and R. Miller. "Geology of the Eastern North Cascades." In *Geologic Guidebook for Washington and Adjacent Areas*. Washington Division of Geology and Earth Resources Information Circular 86, edited by Nancy L. Joseph and others. Olympia: Washington State Department of Natural Resources, 1989.

McGroder, Michael F., John I. Garver, and V. Standish Mallory. *Bedrock Geologic Map, Biostratigraphy, and Structure Sections of the Methow Basin, Washington and British Columbia*. Olympia: Washington State Department of Natural Resources, 1990.

Mierendorf, Robert R. *Archeology of the Little Beaver Watershed, North Cascades National Park Service Complex, Whatcom County, Washington*. Sedro Woolley, WA: National Park Service, 2004.

Mierendorf, Robert, and Gerry Cook. *A Guide to People and Places of the Upper Skagit*. Sedro Woolley, WA: North Cascades Institute, 2010.

Miles, John C., ed. *Impressions of the North Cascades*. Seattle: The Mountaineers, 1996.

Miller, Tom, and Harvey Manning. *The North Cascades*. Seattle: The Mountaineers, 1964.

Misch, Peter. "Bedrock Geology of the North Cascades." In *Geological Excursions in the Pacific Northwest*, edited by E. H. Brown and R. C. Ellis. Bellingham, WA: Department of Geology, Western Washington University, 1977.

Nelson, Don. *Highway 20 at 40: Still Celebrating*. Methow Valley News Online, 2012. issu.com/methowvalleynews/docs/publicservice

Nelson, Jim, and Peter Potterfield. *Selected Climbs in the Cascades*. Vol. 1. Seattle: The Mountaineers, 2003.

Pelto, Maurice S. North Cascade Glacier Climate Project, www.nichols.edu/departments/glacier/.

Pitzer, Paul C. *Building the Skagit*. Seattle: Seattle City Light, 2001.

Pojar, Jim, and Andy MacKinnon. *Plants of the Pacific Northwest Coast*. Vancouver, BC: Lone Pine, 1994.

Rabkin, Sarah Juniper. "Eyes of the World." In *The Way of Natural History*, edited by T. L. Fleischner. San Antonio, TX: Trinity University Press, 2011.

Riedel, Jon L. *Early Fraser Glacial History of the Skagit Valley, Washington*. Boulder, CO: Geological Society of America, 2007.

Riedel, Jon L., et al. *Glacier Monitoring Program*, National Park Service, 2012, at www.nps.gov/noca/naturescience/glacial-mass-balance1.htm.

Riedel, Jon L., R. A. Haugerud, and J. J. Clague. "Geomorphology of a Cordilleran Ice Sheet Drainage Network through Breached Divides in the North Cascades Mountains of Washington and British Columbia." *Geomorphology* 91, nos. 1–2 (2007): 1–18.

Riedel, Jon L., and Michael Larrabee. *North Cascades National Park Complex Annual Glacier Mass Balance Monitoring Report, Water Year 2009*. Sedro Woolley, WA: National Park Service, 2011.

Roe, JoAnn. *North Cascades Highway*. Bellingham, WA: Montevista Press, 1997.

———. *The North Cascadians*. Seattle: Madrona, 1980.

Sanders, Scott Russell. "Mind in the Forest." *Orion Magazine* (November–December 2009), www.orionmagazine.org/index.php/mag/issue/5087/.

Skagit Land Trust. *Natural Skagit: A Journey from Mountains to Sea*. Mount Vernon, WA: Skagit Land Trust, 2008.

Snyder, Gary. *Riprap and Cold Mountain Poems*. Berkeley, CA: Counterpoint, 1965.

Spooner, Alecia M., Linda B. Brubaker, and Franklin F. Foit Jr. *Thunder Lake: A Lake Sediment Record of Holocene Vegetation and Climate History in North Cascades National Park Service Complex, Washington*. Sedro Woolley, WA: National Park Service, 2007.

Stamper, Marcy. *High Hopes and Deep Snows: How Mining Spurred Development of the Methow Valley*. Winthrop, WA: Shafer Historical Museum, 2006.

Suiter, John. *Poets on the Peaks*. New York: Counterpoint, 2002.

Tabor, R., and R. Haugerud. *Geology of the North Cascades*. Seattle: The Mountaineers, 1999.

Thompson, Erwin N. *North Cascades N.P., Ross Lake N.R.A., and Lake Chelan N.R.A.: History Basic Data*. Washington, DC: National Park Service, 1970, www.nps.gov/history/history/online%5Fbooks/noca/hbd/.

Van Pelt, Robert. *Identifying Mature and Old Forests*. Olympia: Washington State Department of Natural Resources, 2007.

Washington State Department of Transportation. Annual traffic report, 1960 to the present, www.wsdot.wa.gov/mapsdata/travel/annualtrafficreport.htm. Traffic reports for the entire State Highway System are available on this website.

———. "SR 20—North Cascades Highway," Washington State Department of Transportation, n.d., www.wsdot.wa.gov/Traffic/Passes/NorthCascades/default.htm.

Wintzer, Niki E. "Deformational Episodes Recorded in the Skagit Gneiss Complex, North Cascades, Washington, USA." *Journal of Structural Geology* 42 (2012): 127–39.

———. "Structure of the Skagit Gneiss Complex in Diablo Lake Area, North Cascades, WA." Master's thesis, paper 3347. San Jose State University, California, 2009.

Photography Notes

MOST PHOTOS IN THIS BOOK were shot with a Nikon D80 and Nikkor 28-200mm VR lens. Though it's not the fastest lens (f3.5), it's clear and steady, and the range is perfect for landscape photography. I had to replace it once after a slip on wet rocks (the sound of glass crunching when I rotated the lens was an unpleasant moment). I rarely use a tripod. I keep my lens clean. I process in a photo-editing program. In the old days we called this working in the darkroom, though of course it wasn't as easy to adjust the images. We manipulate images before clicking the shutter by choosing a telephoto lens or polarizing filter, or adjusting f-stops for depth of field, or making long-exposure water or nighttime images, all things our eyes can't do. The camera is a piece of hardware that will never capture precisely what our eyes see and our brains decode. Having said that, for nature photography I like to keep fidelity with what we see, so I don't like to oversaturate for dramatic color or adjust lighting to something the natural world would not produce. Every photographer interprets what he or she sees by means of choices in subject, equipment, exposure, composition, and postproduction. That is the art in photography.

This project started after almost two decades of doing very little photography. The world went digital. I could take as many pictures as I wanted! And see instant results! Quantity doesn't replace quality, though, so it's still important to be able to take advantage of lighting, to know what your equipment can do, and to have patience and foresight (and to bring mosquito netting). Probably as many photos in this book were the result of serendipitous moments as of planned lighting or composition. Mountain light is dramatic or subtle and never the same twice. I've gone back to the same place, same date, same time of day, and have never captured the same photo. That's part of the adventure. That and the mountain air—it's easy to be patient in peaceful places.

Jack Mountain

Mount Watson

McLeod Mountain

Credits

PAGE V: Epigraph from Scott Russell Sanders, in *The Way of Natural History*, ed. Thomas Lowe Fleishner (San Antonio, TX: Trinity University Press, 2011).

PAGE XI: Epigraph from Saul Weisberg, *Impressions of the North Cascades* (Seattle: The Mountaineers, 1996).

PAGE 3: Epigraph from Bob Mierendorf, *Impressions of the North Cascades* (Seattle: The Mountaineers, 1996).

PAGE 5: Epigraph from United States Geological Survey online glossary, http://geomaps.wr.usgs.gov/parks/misc/glossaryt.html; also available at www.nature.nps.gov/geology/usgsnps/misc/glossaryStoZ.html#T.

PAGE 5: Map from the Washington Department of Geology and Earth Resources.

PAGE 7: Epigraph from Jon Riedel, *Weekend Window: North Cascades National Park,* ABC News video, May 2010.

PAGE 12: Epigraph from Scott Russell Sanders, in *The Way of Natural History*, ed. Thomas Lowe Fleishner (San Antonio, TX: Trinity University Press, 2011).

PAGE 18: Photos of landslide courtesy of the Washington State Department of Transportation.

PAGE 20: Photo of Ross Dam in the 1940s courtesy of Seattle City Light.

PAGE 21: Photo of the gorge crew in 1935 courtesy of the Seattle Municipal Archives, no. 17614.

PAGE 30: Photo of the upper Skagit River valley before Ross Dam courtesy of the National Park Service.

PAGE 31: Photo of five-thousand-year-old hearth courtesy of the National Park Service.

PAGE 53: Photo of McGregor Mountain fire lookout courtesy of the National Park Service.

PAGE 65: Photo of howitzer courtesy of the Washington State Department of Transportation.

PAGE 69: Epigraph from Geof Childs, *Stone Palaces* (Seattle: The Mountaineers, 2000).

PAGE 69: News clipping from 1936 courtesy of the *Spokesman-Review,* Spokane, WA.

PAGE 74: Geologic map of North America courtesy of the U.S. Geological Survey.

PAGE 75: Map of the major geologic domains crossed by the North Cascades Highway courtesy of the U.S. Geological Survey.

PAGE 75: Geologic map of Washington State courtesy of the Washington Division of Geology and Earth Resources.

PAGE 76: Epigraph from Robert Jaffe, "As Time Goes By," *Natural History* 115, no. 8 (2006).

PAGE 80: Epigraph from James Martin, *North Cascades Crest* (Seattle: Sasquatch Books, 1999).

PAGE 80: Plate tectonics illustration courtesy of the U.S. Geological Survey.

PAGE 81: Epigraph from Colin Fletcher, *The Man Who Walked through Time* (New York: Vintage Books, 1967).

PAGE 82: Epigraph from Robert Michael Pyle, *Wintergreen* (Seattle: Sasquatch Books, 2001).

PAGE 89: Photo of crinoid fossil courtesy of George Mustoe.

PAGE 91: Photo of volcanic-gas collection at Mount Baker by David Tucker.

PAGE 92: Epigraph from Harold Bradley, "The Northern Cascades: A Masterpiece to Preserve," *Sierra Club Bulletin* 43, no. 9, (1958).

PAGE 92: Epigraph from Daniel B. Botkin, "Discordant Harmonies," *Nature* 173 (1990).

PAGE 96: Song lyrics reprinted courtesy of John Dillon and Steve Cash.

Index

Italicized page references refer to illustrations.